Tao

Of

ELECTRICITY

http://youtube.com/c/DrChandraSkeekhar
www.ulslab.blogspot.com

FROM SUBATOMIC PARTICLES JOURNEY
TOWARDS UNKNOWN

**For further knowing on
philosophy of Dr Chandra please
visit**

Author Dr Chandra shekhar bhatt
http://youtube.com/c/DrChandraSkeekhar
www.ulslab.blogspot.com

CONTENTS

WAY TOWARDS PERSONAL SAFETY AND SOCIAL RESPONSIBILITY

FROM SUBATOMIC PARTICLES JOURNEY TOWARDS UNKNOWN

"Truth goes beyond three phases. Since creation is infinite souls are infinite. After finishing their part in drama they go back to the source infinite, going and coming is also infinite. The one who can remove the illusion by logic and love of perfect master who has travelled beyond mind and matter can know and remain silent. Researchers and innovators can go on in non ending quest in infinite." Thoughts flowing... can I utter my name? Who am I?

INTRODUCTION

These thought processes came into tangible experiences at the time of my graduation or gradual increasing of my awareness or un folding of my being at Naval dockyard apprentice system in marine electro technology. I met Shalini desai a radioisotope scientist, she had unusual experiences and ways of thinking which a discovery was itself.

When she was tracing the isotope in her instrument she thought what am I doing, wasting my time why shouldn't I trace myself in this vast universe where am I moving to totally uncertain not knowing where I come from and where I have to go, who am I?

She told me that god is like electricity it is flowing you may feel it or you may not feel it. One of my friends told me so your source is woman. I said yes in general they are purer at heart, so they go faster and higher in meditation. Monk Bodhidharma the spreader of martial arts from India to china was initiated by a female. Grandmaster Yip Man the master of Bruce lee learnt Win chun from a female Ng mui who discovered Win chun Kung fu. This book is mainly for innovators and artists. I would put art or artist in an artistic way of searching by Robert Edmund Jones, a visionary designer and director. He would begin his first lecture to his students by slowly, silently looking at each and every one, back and forth, almost with the pacing, inevitable rhythm of a lion in a cage. And then suddenly he would cry, "I am studying you very carefully because I know here in this room there are a few, only a few who are doom eager...doom eager to be an artist. And the artist is doom eager, but never chooses his fate. He is chosen and anointed and caught.

GLOSSARY

1)   Aashirwaad Blessing.

2)   Abhyasa   Practice.

3)   Abhinivesah   Instinctive clinging to life or fear of death.

4)   Adhara   A support.

5)   Agama   Evidence of an acceptable authority.

6)   Ahamkara   Ego.

7)   Ahimsa   Non-violence.

8)   Ajna-chakra   The Chakra situated between the eyebrows.

9)   Alasya   Idleness, laziness.

10)   Anahata-chakra   The chakra situated in the cardiac region.

11)   Anga   The part.

12)   Anumana   An inference.

13)   Antar Inside.

14)   Apana   Vital air which controls the function of   elimination of faeces.

15)   Ardha   Half.

9

16) Astanga Yoga   Eight parts of the Yoga.

17) Asana   Posture.

18) Asta Eight.

19) Asmita   "I"ness or Egotism.

20) Asteya   Non-stealing.

21) Asvini-mudra   The contraction of the anal sphincter.

22) Atma Soul.

23) Aum The primordial sound.

24) Avidhya   Ignorance.

25) Avastha   State.

26) Ayurveda   Traditional Indian system of medicine.

27) Bandha   In a posture where certain parts of the body
    are contracted or locked or controlled.

28) Bahya   Outside.

29) Bhastrika   Pranayama where air is forcibly draw in and
    out as bellows used in a furnace.

30) Bhoga   Enjoyment.

31) Bhramari   Pranayama where during exhalation a soft

humming sound like the murmuring of a bee is made.

32)   Bhrumadhya   The place between the eyebrows.

33)   Bhujangasana   Cobra pose or posture.

34   Brahamarandhra An aperture in the crown of the head through which   the soul is said to leave the
body..

35)   Chakra   Centre of vital channels in the etheric body.

36)   Chandra   Moon.

37)   Chakrasana The wheel posture.

38)   Chitta   Consciousness.

39)   Darshan   Sight of a guru or deity.

40)   Dhyana   Meditation.

41)   Dharana   Complete attention.

42)   Dhanurasana   Bow posture.

43)   Dukha   Sorrow.

44) Dvesha   Hate or dislike.

45) Eka   One.

46) Ekagrata   One pointedness.

47) Ekapada   One leg.

48) Gow mudra Cow posture.

49) Guru Dispeller of the darkness.

50) Guna Constituents of nature.

51) Hatha Yoga 'Ha' means the sun and 'tha' means the moon. Hatha Yoga is   to harmonize the solar and lunar energy.

52) Hanuman asana   Posture named after Hanuman.
                    The disciple of Lord Rama.

53) Halasana   The plough pose.

54) Indriyas   Sense organs.

55) Isvara or Eshvana God.

56) Jalandharabandha   Chin Lock.

57) Janu   The knee.

58) Janusirsasana   The posture where head touches to the knee.

59) Japa   The repetition of the Mantra.

60) Jivatma   The individual soul.

61) Jnana Knowledge.

62   Kaivalya   Final emancipation.

63) Kapalbhatti Forehead brightener. 64)
Pitaji Father

SOURCES

Path of the masters

Autobiography of a yogi

Jonathon Livingston seagull

Teachings of Don Juan

Vedas and Upanishads

Kitab e Mirdad

Bible

Bhagavad-Gita

Books of Radhasoami Masters

Six systems of Indian Philosophy

Fundamentals of Indian and western
philosophy

Sufi philosophy

Naval Docks electrical knowing

Anurag Sagar

ACKNOWLEDGEMENT

I Acknowledge and respect the genius of all those who had touched my being and we happen to exchange our views globally.

It is impossible to mention the names of all philosophers. Scientists, performing artists, workers from all walks of life, who have become instrumental in helping this name and form named shekhar, grow to know the formless.

I owe special thanks to PratapAntony,Pradeepkumar,Pradeepkamat,Vijayrane,Venkateshraghvan and family, Rajat pandey and family.   Hoshi daruwalla, Darshana Doshi and Manibhai Doshi a close associate of Shri Subhash Chandra Bose and founder of Rotary club chain at south Bombay for giving help and inspiration A Hillman from navy named pathak Arun pradhan and family, Ujjwal roy, Mahesh jabbar. A Chinese friend and fellow practitioner Yu chu Liang who helped Vajramukti grow My friend Mr. Fung and Tong from Oberoi Sheraton hotel and all my colleagues at that time when I was working to support my family. Advocate Subodh Desai & his family Subodh father Justice Desai and his mother were great inspiration for me, Subodh aunty Shalini Desai who inspired me for real education. Late Rami chand Rajbar Mani Chand Rajbar from chand dynasty who helped us at the time of crisis to literate our sisters.

My sincerest thanks to Late Dr Ram Bhosle ,Gagangiri Maharaj and Swami Chakrajit or Ghateji pray their soul rest at peace. Sitting with them I could touch the personality like Joseph Stalin and Deathless guru and many more. I can never forget the love in their eyes who came to meet me when I got initiated at Indian naval dock I Sincerely thank Shyamdutt and family, Rahul Hindoliya, Sudeep lama, Mr. Negi, Alok rana, Jaman singh Rawal and Inder singh Rawal, Late Bharat chand chacha and family.

My dear students and friends who wanted my work to be published Ajey dambal, sheffy George, Abhishek ghosh, , Shaun cooper, Vimal kumar choubey, Unni Krishnan Nair, Surendra singh Rawat James, Ajay, Dharampal singh Rana, BalbirsinghTatwalia, Jasvinder from bank star technology, Malik the great from Haryana, Late Shri Vitthal Bhatia, Dr Dallal, Venkatesh Raghavan, Mathew, Mighty simple hearted Billa, Rai, charge man Visvekar all my batch mates and our dear Instructor Shri Bisht. Our overall in charge who advocated humanness by increasing different other attributes such as sports and other arts pertaining to cultural activities and removed ragging, also helped boys coming from different parts of the country to come out with their own ideas, he was great Commander Gildhiyal .

My gratitude for the following people cannot be adequately expressed. My yoga teachers Shri Dilip vagchaure, Alkatai and Usha tai, My Martial arts teachers Sensei Leslie Fernandez and Sensei Oliver Fernandez.

Words are inappropriate to thank my wife Shobha and her family without their support this was not possible.

Dr Chandra Shekhar sitting at the feet's on right side of Martial arts legend Dr Ram Bhosle the descendent of Shri Chattrapati Shiwaji Raja Bhosle student of Sir Herbert Barker of United Kingdom seated in between. On left side of Dr Ram is Michael Trembath who wrote a book on Dr Ram "The hand that heals" inspired by Dr Chandra Shekhar and Pradeep kamat. Dr Ram was very keen for the work on self enquiry and electricity to be published. He used to say Chandra Shekhar actually first aero plane was tested at this land of Marathas on the basis of vimanshastra of Bharadvaja muni but the man who did had disappeared.

Preface

It is with great pleasure. I am writing this Preface to the book titled "Tao of Electricity" by Dr Chandra Shekhar

The life of Dr Shekhar is nothing but a Saga of struggle and sacrifice. He has been working after his ninth class to support his father and family

It will be experienced by reading the pages further This is the story of a single mans quest for mastery in his own being with all the problems of life Dr Shekhar elevates the way of Artist and encourages individuals to bring out the artist from their beings. His accomplishments in the field of yoga and martial arts are legendary. He tries to remove the myth from the traditions by increasing the awareness and touching the consciousness of the individuals. This book is based on his search for the path while he was being trained on marine electricity with Indian Naval shipyards and remains of the philosophical talks between Chandra Shekhar and his father.

His father wanted these talks to be published after his death. Chandra Shekhar is doing it so. This book way of electricity actually was a means for his search for the ultimate; he has offered his healing touch by means of teaching Yogasanas with his quantum of yoga meditation and martial arts to families like Kamani group of Industries. The family who has been bestowed mysterious touch by Ramjibhai Kamani who used to meditate regularly for knowing thyself. The material richness was the outcome of his past karmas performed. It will serve as a tool for philosopher thinkers innovators and mariners to become a wholesome human being. It is a pleasure to go through the book which makes us aware and help us to grow in our own path by Tao of electricity.

By Manibhai Doshi

           Close associate of Netaji Subhash Chandra Bose

           Founder of chain of Rotary club at south Mumbai

*It was all by his grace
complete Master Anami
beyond names and forms as
described by Lord Kabir in
his ocean of love*

Way of electricity

When I was in fourth class of primary conventional school, one of my uncles died. In my innermost core I was thinking I will not die and I was sure about that and thoughts were there with interactions with my father which I will be giving in my books or works, how can I accept death when I have not experienced death.

These thought processes came into tangible experiences at the time of my graduation or gradual increasing of my awareness in Naval dockyard apprenticeship in electro technology, I met Shalini Desai a radioisotope scientist, she had unusual experiences and ways of thinking which a discovery was itself.

When she was tracing the isotope in her instrument she thought what am I doing, wasting my time why shouldn't I trace myself in this vast universe where am I moving to totally uncertain not knowing where I come from and where I have to go and who am I.

She told me that god is like electricity it is flowing you may feel it or you may not feel it.

It all depends on your awareness levels. Once it happened so when I was reading Sama Veda deep inside I was speaking to my creator that I will definitely search you where ever you are, all of a sudden as if air spoke to me the sound came in the form of the language "Mujhe kahan dhundaga mai to har kan main samaya hoon"

Where you will search me I am inside every particle.

As the bulb goes off, a technician fixes it or changes it, same is with human form, a true yogi can make changes in the cosmos and make even dead men alive but I couldn't Accept the idea that time I was aware that electricity is the finest and subtlest

Achievement of science towards the subtler levels, from those days when it was believed that sun revolves around the earth.

The Reformation was led by German reformers which resulted as emancipation of the state from clerical control. Initially the state was universally regained as subordinate to the church, the revolt which begin on grounds of Religious dogma & practices almost immediately begin to take on political and social overtones. The Reformation removed Germany from the dominant influence of the catholic church, liberated the secular German princes from the financial burdens imposed by Rome & gave them the wealth of the church with which to strengthen their regime, and ultimately resulted in the civil regulation of religion and education. Thus the result was separation of church and state. Religion, science, philosophy and art now were on their own ways.

For more than fourteen centuries a system devised by Claudius Ptolemy, second century Egyptian astronomer, was accepted as the true conception of the universe. Ptolemy held that world is fixed & immovable sphere, situated at the center of the universe, about which all celestial bodies including sun and the fixed stars revolve. The Ptolemaic structure remained intact until the coming of the era of intellectual awakening in Europe the Renaissance. Then the revolutionary work of a Polish astronomer & churchman Nicklaus Copernicus 1540 he attempted to demonstrate that the earth is not stationary but rotates on its axis once daily & travels around the sun once each year. The sun was inert & passive, stationary amid revolving planets. Its only functions were to supply light & heat.

The universe was strictly limited. Outside the sphere of the stars, as Ptolemy had taught, space ceased to exist. Neither did Copernicus abandon Ptolemy's system of epicycles, reaffirming the ancient astronomer's assertion that for each of the orbits there was a diff center. These features of Copernicus system were to be corrected by later astronomers.

Sir Francis Bacon's doctrines 1605 were first set forth in detail in Proficiency & Advancement of learning, he did much to usher into the world the present scientific age. Bacon firmly asserts that the right method of scientific research, that is inductive method, could give mankind sovereignty over nature.

Learning is divided by bacon into three categories, History which replenishes the memory Poetry which gives new dimension to imagination & Philosophy which probes the mind to reason.

Johannes Kepler was born a little more than a generation after the death of Copernicus. One of the great creative scientists Kepler represents both the mysticism of the middle ages & the positive reasoning spirit of the Renaissance.

Though German born Kepler's investigation in astronomy were carried on principally in Prague, when he had gone as a young man to serve as Tycho Brahe's assistant. When Brahe died in 1601, Kepler became successor, to his patron's post. The position was that of court astrologer, with Kepler's professional duties limited to testing horoscopes and issuing prognostications.

Of immeasurable value to his scientific investigation was the fact that he inherited Tycho Brahe's voluminous unpublished collection of astronomical data. His records covered a period of thirty five years and were indispensible to Kepler in carrying forward his own studies.

It had been universally believed that the planets move in circular paths a theory accepted unquestioningly even by experiences Kepler soon came to realize , however that there were striking discrepancies between the circular orbit theory of planetary motion & the observed facts. Six planets Mercury, Venus, Earth, Mars, and Jupiter & Saturn were then known. Observations had revealed that these planets move in their courses at variable speeds, their rate decreasing as their distance from the sun increases.

Kepler begin testing variety of hypotheses, & making innumerable calculations, he finally developed two epoch making laws. In his first published "The new astronomy" According to the 1st law the planetary orbits are not circular, but elliptical, the sun occupying one of the ellipses. With this law Kepler upset the old Aristotelian notion that the circle is a perfect figure & therefore the planetary orbits must be circular.

Kepler's second law is concerned with the fact that the speed of a planet varies as it approaches or recedes from the sun, observed variation in speed also such that planets sweep over equal areas in equal time. If one imaginary line is drawn from the sun to a planet moving in its elliptical path, & one end of the line is considered as fixed at the sun while the other end moves with the planet, the quasi-triangular areas swept over by the line in like periods of time are always the same.

The third law is in mathematical terms. The squares of any two planet's periods of revolution about the sun are proportional to the cubes of their mean distances from the sun. Although he suspected that the sun exercises some physical control over the planets but he being religious in nature believed that each planet was being held in its course by the governing power of a guiding angel. The main idea of putting so many thoughts in this one book that too technical or techno philosophical is Indian upbringing, we are so much involved in religious and philosophical talks in our houses and the prayers we do in temples that is not in line with the science which we study at schools. There should be books incongruence with philosophy and technology so that we don't make machines out of human beings. At the same time we are aware of the facts. Also so called science should not become conditioning to disrupt the eco friendliness. Education should not become literary accumulation. To explain further I would give an example. A boy trained with Indian navy after his tenth standard for four years. He went to college for conventional accumulation that is graduation. The Dean asked him have you completed ten plus two. The boy said I have completed ten plus four. The dean cannot understand the boy's explanation so the boy left. The system became to make high level clerks. My father wanted to understand further so I explained him by giving example of what had happened at America. After successful launching of Sputnik by Soviets in 1957 education became competitive in America in comparison with soviet for technological superiority. School became more competitive and students were increasingly grouped by ability into college & choices were made by guiding counselors & choosing was among the given options. If such conditions comes into existence. What is the idea behind? Growing up professionals, where goes the essence and is that education? Writers like Paul Goodman in compulsory mis-education. The Community of scholars, growing up Absurd & Edgas Friedenberg in The Vanishing Adolescent argued that Children's were being schooled rather than educating. These creative writers of America contributed their awareness to the public. Then Americans came up with the idea of Alternative school means any school that provides alternative learning experiences to those provided by the conventional school and that is available by choice to every family within its community at no extra cost. University degree can make good literate clerks or high level clerks but not innovators. It doesn't help people to grow as buds transforms into flower. Take example of Great Srinivasa Ramanujam he was largely self educated. When he was 16, he found an 1856 book by G S Carr[C], which listed theorems and formulae and some short proofs'. Using the tutorial textbook he taught himself mathematics and by the age of 17 was engaged in deep mathematical research, studying Bernoulli numbers and divergent series and calculating the Eular-Mascheroni constant to 15 decimals he didn't graduate in conventional way, as I worked with my brother in law on thesis to put the juxtaposition in order. I found the work they were doing was just to get a job. But he is not to be blamed it is the system which gradually has to undergo reformation. This can be done by every individual contributing some of his effort for his country. My brother in law was swollen with ego that he was msc first class. His second class batch mate was selected by Germany for research work and now he works there. Such people becoming professors where they can lead the country just mugging. But they are not to be blamed it is the system which needs reformation, it will happen slowly by making people aware. There are good people also like left their job and are working to educate countrymen's but from these kinds of people very less can retain their own way of unfolding. Like how Master Bruce lee started his classes at America. Giving a title In the name of man who was once fluid. Pointing out the conditioning and removing the enshrouding given by classical mess. These peoples always had tough time but they are the one who build the nation, as Shri Chanakya did by removing Nand Dynasty and giving power to Morya dynasty. Country went till his grandson Emperor Asoka. Being religious in nature Kepler suspected that the sun exercises some physical control over the planets but he being religious in nature believed that each planet was being held in its course by the governing power of a guiding angel.

The invention of telescope was done on 1609 by a Dutch man Johannes Lippershey by the aid of which visible objects, even though far distant seem as if near at hand.

Galileo himself prepared a tube of lead in the ends of which he fitted two glass lenses, both plane on one side, one being spherically convex, the other concave on the other side. He was the first astronomer to use it in order to observe the heavenly bodies.

Continuing his observation Galileo discovered sunspots & by watching their changing positions calculated that the sun revolves on its axis every twenty-four days. He found that the Milky Way was "nothing else but a mass of innumerable stars planted together in clusters," there were four moons revolving around the planet Jupiter. Further observation revealed the phases of Venus & the rings surrounding the planet Saturn. "We are absolutely compelled to say asserted Galileo, "that Venus & mercury also revolve around the sun, as do also all the rest of the planets.

Galileo's greatest gift was inductive method through which our knowledge of the world has grown to be a million times greater than that of the ancients. Galileo was the son of Pisan nobleman. The scientific revolution, beginning in the sixteenth century with Copernicus, hastened the process of disillusionment & it was to this new world of science that Descartes gravitated. He created co-ordinate geometry, by uniting algebra with geometry for discovering facts about nature, to study the changing patterns in the nature and nature as a whole. He found that older math was lacking and he created co-ordinate geometry for his work. He had a deep philosophical awareness for inner journey and believed that pineal gland is the seat of the soul as Hindus believe between the eyebrows is the third eye of Lord Shiva. He concluded that mathematical method was the ideal tool to apply in every sphere of knowledge.

Rufus Sutter refuted saying "the technique of pure mathematics is not enough when the thinkers seek to decode the laws of other sciences the thinker must quit his arm chair & go into the laboratory, on the other hand mathematical physicists, such as Albert Einstein exemplify Descartes method, depending upon little else besides pencil & paper to carry on their research.

Rene Descartes used the method of doubt to find absolutely certain starting point for the building up of our knowledge [Cogito Ergo Sum] to doubt is to think, to think is to be.

The great scientist Isaac Newton proved that the Earth, revolving round the sun, follows the same laws of mechanics as the moon to revolve around the earth & the other planet around the sun, too. The law of universal gravitation proves that the earth and celestial bodies not only of our galaxy but of all constellations are bound together, & thus constitute as simple unified system that is our world or the universe. Celestial bodies consist of same elements as does the earth. The same common basic element had been discovered in all other bodies of the universe as on the earth. The main ingredient of meteorites, for ex which comes from the depths of space is iron, an element very common on earth. This is convincing proof that there is nothing immaterial.

Newton believed that space & time exist independently of matter and material things. He concluded that space is absolute i.e. independent of matter. The geometrical proportion of space, he supposed, are same in all directions. His view was of materialism.

Lenin attached great importance to the idea that space and time exist objectively, because it is directed against the subjective idealistic view of space and time. This view derives from the 18th century English Philosopher Hume & German philosopher Kant according to both of them time and space have no objective content and modern idealist try to falsify some of the vital discoveries of modern science, in particular of physics, in order to revive this subjective idealistic view for example, the theory of relativity one of the key advances of 20th century bended science to this end. Einstein was very much attracted to German philosopher Kant.

Einstein's theory of relativity showed that there is no unified, unchangeable, Newtonian space. The properties of space change, they are dependent to material things. For instance, it was found that the length of a body decreases as its speed increases. Let us imagine that a train is rushing past a station platform with a velocity close to that of light. We should naturally suppose that the length of the platform as measured by the engine driver & someone on the platform would be the same. By doing the precise mathematical calculation based on the theory of relativity it shows that it is not same. The driver will find that the platform has decreased in size, while the person at the platform will find that the moving train has decreased in size, this isn't the optical illusion but objective fact that Space is relative.

The same applies with time as the speed of a material system increases, time in it slows down. Time also slows down in a very strong gravitational field. If a future spaceman is put in a space orbit, the flow of time into the spaceman's ship will be much slower than on the earth he has left behind. On returning to earth, his peoples whom he left behind will be older than him.

So Einstein's theory of relativity has proved that space & time are not absolute in the Newtonians way, because they are not, as Newton thought, the same unchangeable things throughout the whole of the universe. But space & time are absolute in the philosophical sense; everything in the world has spatial properties & exists in time, nothing has existence outside of time and space. But space & time are relative in physical sense, for they depend on the properties of moving matter. Matter, space are bound together. Matter in essence is particles. The same particles were further broken down to sub atomic particles. Which resulted in invention of electrons, remarkable work was done by Thomas Edison on electricity. Vibrant touch on electricity was given by Nikola Tesla. Nikola Tesla was an artist Philosopher and Prophet and Scientist of vibrant age of electricity.

He made a revolution to take over from Direct current DC and made Alternating current power generation and transmission possible. Because of the direct and alternating current in conjunction working we got other tangible inventions in the field of Radio, Television, telephone, auto ignition and many other avenues possible. Let's have some idea of what is d c the full form is direct current its produced by a machine or a generator known as dc machines

An electric Machine that converts Mechanical energy into electrical energy

Principle of Operation

An electric generator is based on the principle that whenever a magnetic Flux is cut by a conductor an electromotive force is induced which will cause a current to flow if the conductor circuit is closed.

Therefore, the essential components of a generator are

(a) A magnetic field

(b) Conductor or a group of conductors.

(c) Motion of conductor in conjunction with Magnetic field.

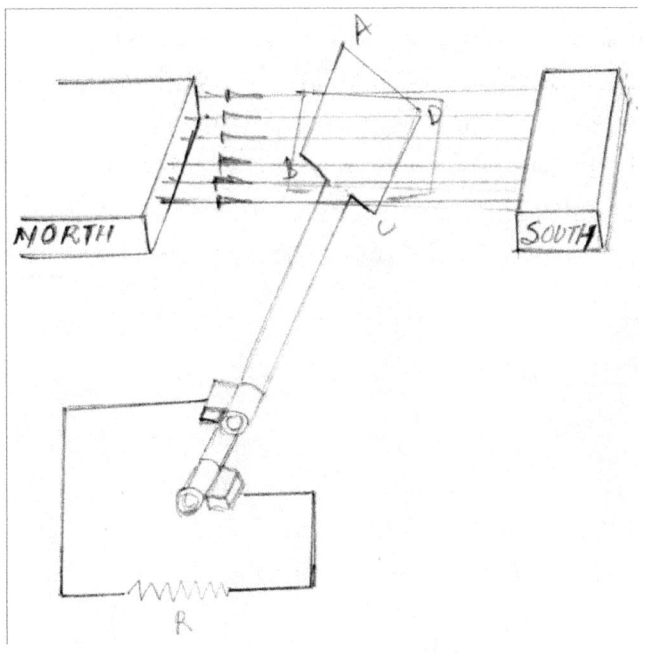

NORTH

SOUTH

A

B

C

D

R

Consider a Single turn Loop conductor ABCD rotating clockwise in a uniform Magnetic field with a constant speed. As the Loop rotates the flux linking the coil sides AB and CD changes continuously. Hence the electromotive force induced in these coil sides also changes.

When the loop is in position 1 – electromotive force generated is zero as coil sides AB & CD are not cutting Flux.

When the loop is in position 2 – the coil sides are moving at an angle to the flux , therefore a low electromotive force is generated.

When the loop is in position 3 – the coil sides AB &CD are at right angle to the flux and are cutting flux at maximum rate. Hence at this instant the generated electromotive force is maximum

At position 4 – the generated electromotive force is less because the coil sides are cutting the flux at an angle.

At position 5 – no magnetic lines are cut and hence, the induced electromotive force is zero.

At position 6 – the coil sides move under the pole of opposite polarity and hence the direction of generated electromotive force is reversed.

At position 7- the maximum electromotive force is generated in the reverse direction

At position 8 – the flux in the reverse direction again reduces

And again at position 1 – induced electromotive force being zero the cycle repeats itself with each revolution of the coil.

Please Note: The electro motive force generated in the coil loop AB, CD is an alternating one. It is because any coil side, say AB, has electro motive force induced in one direction under the influence of N-Pole and in the other direction under the influence of South Pole.

The alternating Voltage thus generated in the Loop can be converted into direct voltage by a simple device called "commutator". Commutator is a kind of mechanical rectifier

A rotating Commutator is in contact with stationary carbon brushes and lead current to external load. With this arrangement, the commutator at all times connects the coil side under S-Pole to the positive Brush and that Under N- Pole to the negative Brush.

Thus the alternating voltage generated in the armature of the DC generator is converted into Direct Voltage by the commutator. The Sole purpose of the Carbon brushes is simply to lead the current from the rotating Coil Loop or Winding to the external load.

A pulsating direct voltage is produced by a single loop of coil as shown in the fig. and is not suitable for commercial use. What we require is a steady direct voltage. This can be achieved by using a large number of coils connected in series. The resulting arrangement is known as armature winding

The DC Generators and Motors have the same general construction. Any DC Generator can be run as a DC motor and Vice versa. All DC Machines have five principle components:

Armature core

Armature Winding

Commutator

Field System

Brushes

Field System

The function of the field system is to produce uniform magnetic field within which the armature rotates. It consist a number of poles bolted to inside of circular frame called yoke. The yoke is usually made of solid cast steel where as the pole pieces are composed of stacked laminations. Field coils are mounted on the poles and carry the DC exciting current. The field coils are connected in such a way that adjacent poles have opposite polarity.

The magneto motive force developed by the field coils produces magnetic flux that passes through the pole pieces, the air gap, the armature and the frame.

The air gaps should be limited to 0.5mm to 1.5mm. Since the armature materials are composed of materials that have high permeability, most of mmf of field coils is required to set up flux in the air gap. Thus by reducing the air gap we can reduce the size of field coils.

Armature Core: The Armature core is keyed to the machine shaft and rotates between the field poles. It consists of slotted soft iron laminations (about 0.4 to 0.6mm thick) that are stacked to form a cylindrical core. The laminations are individually coated with a thin insulating film so that they do not come under electrical contact with each other. The purpose of laminating the core is to reduce the eddy current losses. The laminations are slotted to accommodate and provide security to the armature winding and to shorten the air gap for the flux to cross between the pole face and the armature.

Armature Winding

 The slots of the armature core hold insulated conductors that are connected in a suitable manner. This is known as armature winding. This is the winding in which electromotive force is induced.

The armature conductors are connected in series so as to increase the voltage output and parallel path arrangement to increase the current output

The armature winding of a DC machine is a closed circuit winding, the conductors being connected in a symmetrical manner forming a closed loop or series of closed loops

Commutator

A commutator is a mechanical rectifier which converts the alternating voltage generated in the armature winding into direct voltage across the brushes. The commutator is made up of copper segments insulated from each other by Mica sheets and mounted on the shaft of the machine. The armature conductors are soldered to the commutator segments in a suitable manner to form armature winding.

Commutator should always be well balanced and true on the shaft, because any eccentricity will cause the brushes to jump producing extra sparking. The sparks may burn the brushes and overheat and carbonize commutator.

Depending upon the manner in which the armature conductors are connected to the commutator segments, there are two types of armature winding in a DC machine

Lap winding, Wave winding.

Wave Winding

Types of Armature Winding

The armature coils are connected in series in such a way that electromotive force adds to each other. There are two types of dc armature winding (i) Wave Winding (ii) Lap Winding.

(i) Wave winding: In this arrangement, the armature coils are connected in series through commutator segments in such a way that the armature winding is divided into two parallel paths irrespective of the number of poles of the machine.

Lap Winding

In this arrangement the armature coils are connected in series through commutator segments arranged in such a way that the armature winding is divided into as many parallel paths as the number of poles on the machine.

Armature Resistance

The resistance offered by the armature circuit is known as Armature resistance (Ra) and it includes

 (i) Resistance of Armature windings

(ii) Resistance of Brushes

The armature resistance depends upon the arrangement of windings of the machine. Except for small machines, its value is generally less than 1 ohm.

Brushes

The purpose of brushes is to ensure electrical connection between rotating commutator and stationary external Load circuit. The Brushes are made of carbon and rest on the commutator. The pressures on the brushes are adjusted by means of adjustable springs. The brush pressure must be just right, if it is too large, the resulting friction would cause heating of the commutator and brushes. On the other hand, if it is too weak, the imperfect contact with the commutator would produce sparking

Types of DC Generators

The Magnetic field in a DC generator is produced by electromagnets rather than permanent magnets. Generators are generally classified according to their methods of field excitation:

(i) Separately excited DC Generators

(ii) Self excited DC Generators

The behavior of a DC Generator on load depends upon the method of field excitation adopted

Separately excited DC Generators

A DC Generator whose field magnet winding is supplied from an independent external DC Source is called separately excited Generator.

The voltage output depends upon the speed of rotation of the armature. And field current. Greater the speed and field current, greater is the generated electromotive force

Self Excited DC Generator

A DC Generator whose field magnet winding is supplied from output of the generator itself is called a self excited generator.

There are three types of self excited Generators:

Series Generator

Shunt Generator &

Compound Generator

Series Generator

In a series wound generator the Armature, Field winding and the Load are in series.

The general characteristics of a Series Generator:

The machine has increasing voltage as the load current increases, since the field excitation increases with the current up to its maximum limit which is limited by the following effects:

Magnetic saturation takes place so that any further increase in current causes only a small increase in Flux

The increase of armature current causes increase of armature reaction, which reduces total flux, therefore the generated e.m.f

The terminal voltage is less than the generated e.m.f by the amount of voltage drop in the machine due to load current flowing in series through the armature, field and the brushes.

Shunt Generator

The field windings are connected in parallel with the armature conductors and the load.

The external load characteristics of a shunt generator are:

When Zero load current is flowing, the field is fully excited and induces the open circuit terminal voltage. As the load is placed on the machine, the terminal voltage drops due to the following effects:

Resistance of the armature, brushes etc, causes a voltage drop as current flows.

Armature reaction reduces the total flux and therefore the induced e.m.f

As the terminal voltage falls the current on the shunt field also falls causing further decrease of Flux and of induced electromotive force

Within normal limits of load, the terminal voltage may only fall by 4%, but if the resistance of the external load is further reduced in an attempt to obtain greater loads, the voltage falls at an increasing rate mainly due to the effect of above.

Compound Generator

Compound generator are further classified as follows

Compound Generator

Short Shunt====Cumulative-----Differential

Long Shunt---------Cumulative--Differential

Characteristics of Compound Generators

The compound wound Generators make use of Shunt and series excitation thus combine a raising series characteristic with a falling shunt characteristic. If the increase of voltage between no load and full load due to the series winding is just equal to the decrease of voltage using the shunt winding alone, then the complete characteristic of the machine will be approximately parallel to the current axis and the terminal voltage at Full load will be equal to that of No load. The machine is said to be level compound.

Armature Reaction

It is the effect of magnetic field, produced by the armature current, on the distribution of flux under the main poles of a Generator. The effects of armature reaction are:

It de-magnetizes or weakens the main poles

It cross magnetizes or distorts it.

Effects of Armature reaction

It causes heavy sparking on the commutator.

It weakens the Generated electro motive force

In small machines the armature reaction can be prevented by shifting the brushes towards the direction of the rotation of armature.

But in large machines that undergo violent load fluctuations of load, compensating windings are used for neutralizing the flux due to the armature current. These compensating coils are put in the pole faces and carry current in the opposite direction to the current of the armature.

Explanation of Armature Reaction

When the Generator is loaded, the armature conductor carries current, the current carrying conductor produce magnetic flux of its own which effects against the flux due to main poles. This effect of magneto motive force set up by the armature currents on the distribution of flux under main poles are called armature reaction. The armature reaction is there only when the armature conductors carry current and its effect increases as the armature carry more current.

Commutation

The commutation is defined as the process of collecting current from the commutator in the short circuited coil during its transfer from one armature coil to another. During the period of short circuit by a brush, the current in a short circuited element should be reversed and brought upto its full value in the reversed condition. If the current is not attained its full value in element A and since element B is also carrying full current, the difference between the currents through elements B&A has to jump from the commutator on to the brush in the form of Spark. Thus, one of the causes of sparking at the commutator is the failure of the current in the short-circuited elements to reach the full value in the reversed direction by the end of short circuit.

Reactance voltage

When the current in the short circuited coil is reversed, the rate of change of current is so great that the self-induction of the coil sets up a back electro motive force which opposes the reversal. This induced voltage is known as Reactance voltage.

Methods to reduce sparking due to the reactance voltage:

Sparking on the commutator due to Reactance voltage can be reduced by the following two methods:

EMF commutation: In this method a short circuited coil has a voltage that will neutralize the reactance voltage.

Inter poles : it is a better method of providing the commutating field. These are small auxiliary poles placed in the GNP (Geometric Neutral Plane) that is midway between main poles. The polarity of the inter poles in case of generator must be same as the main pole ahead. These are connected in series with the armature so that the commutating field must be proportional to the armature current. The modern machines are capable of operating between no load & 20~25% overload load with fixed brush positions without appreciable sparking.

Please Note: Inter poles do not reduce armature reaction.

The other method of obtaining good commutation is to use high resistance brushes where it is possible to open two parallel paths between two adjacent elements on the commutator, whereas the brushes having low contact resistance, there is no inducement for the current to take the second path.

Equalizing Rings

The equalizing conductors, which are in the form of Cu rings at the armature back connected to symmetrical points in the armature winding are called equalizing Rings.

The Function of equalizing Rings is to avoid unequal distortion of current at the brushes thereby helping to get spark-less commutation.

No of Rings = $\dfrac{\text{No. of conductors}}{\text{No. of pairs of poles}}$

Reasons for high sparking on the commutator

In addition to effects of Armature reaction and Reactance voltage the sparking may also be due to poor commutation due to the following:

Worn out brushes

Sticky brushes

Wrong grade of brushes

Inadequate pressure on brushes

Brushes not properly bedded

Unequal brush spacing

Rough commutator surface

Projecting intersecting Micas

Incorrect brush position

Earth fault in Armature

Armature short circuit

Reversed inter pole coils

High Ambient temperature

Methods of reducing Sparking in the commutator

To reduce sparking on commutator, following methods are used

Uses of Inter-poles which are auxiliary poles placed between main poles and are connected in series with the armature.

By using high contact resistance brushes such as graphite carbon (commutation)

By using compensating windings on the field which are connected in series with the armature (armature Reaction)

By using short chord winding in the armature

Types of Losses in a DC Machine

The following are the losses observed in a DC Machine:

a) Copper Losses: Those losses which take place in armature winding and the field winding of the DC Machine ($I^2 R$). They are further classified as (i) Armature Cu loss (ii) Shunt Cu loss & (iii) Series Cu loss

b) Iron Losses: Means those losses which occurs in the armature and field core and are sub-divided into (i) Hysteresis loss & (ii) Eddy current loss. These losses are practically constant for all DC Machines. (Losses due to Magnetic effect)

c) Stray Losses/Mechanical Losses: Usually Iron or Core losses along with friction losses are collectively known as stray losses. Friction Losses are those losses which include (i) Brush Friction, (ii) Bearing friction & (iii) rotating armature

Mechanical Input-----Iron & Friction Loss ----
Electrical Power developed ------Copper Loss ---
--Electrical Power Output

Types of Carbon brushes

Carbon Brushes: Amorphous carbon in various forms such as lamp black, Gas carbon or coal is first reduced to fine powder and mixed with suitable binding material such as pitch ot tar and compressed into blocks. To carbonize the whole mass the blocks are baked at high temperature. This process also imparts the required mechanical and electrical characteristics.

If a small quantity of graphite is added to the original carbon before heat treatment the high speed running and the current capacity of the brush are much improved.

They are used in small machines whose rubbing speed, commutation and current density are small.

Natural Graphite Brushes: The thermal conductivity and the lubricating property of graphite are very high. The Graphite brushes are generally suitable for large heavy DC Generators and motors.

Electro-graphitic Brushes:

The material is subjected to a very high temperature under controlled atmospheric conditions. Its effect is to convert the carbon and the binder into almost pure artificial graphite. This class of brush has many characteristics such as

Good Mechanical strength and toughness

Co-efficient of friction is Low.

Capacity to operate even under most arduous condition is satisfactory.

Current carrying capacity is High.

Good High speed running capacity

They are suitable in large DC Machines with the highest peripheral speed and in the most difficult conditions.

Metal Graphite Brushes : Powdered copper or bronze is mixed with graphite and a suitable binder, the whole is molded into blocks under heavy pressure. The metal portion may vary from 50 to 95% giving large current carrying capacity. It also maintains Low friction loss.

These are suitable for use on Low and medium slip Rings and Low voltage DC generators where commutation is not difficult and Low contact resistance is desirable.

Draw backs of direct current generation and activation were taken to a higher state by thinker philosopher mystique Nicola Tesla

Apart from his innovative sense, Tesla had studied Philosophy at the University of Prague. Practicality of his knowing can be grasped by the visions he was getting in his own words.

When I close my eyes I invariably observe first, background of very dark and uniform blue, not unlike the sky on a clear but starless night. In a few seconds this field becomes animated with innumerable scintillating flakes of green, arranged in several layers and advancing towards me. Then there appears, to the right, a beautiful pattern of two systems of parallel and closely spaced lines, at right angles to one another, in all sorts of colors with yellow ,green, and gold predominating. Immediately thereafter, the lines grow brighter and

the whole is thickly sprinkled with dots of twinkling light. This picture moves slowly across the field of vision and in about ten seconds vanishes on the left, leaving behind a ground of rather unpleasant and inert grey until the second phase is reached. Every time, before falling asleep, images of persons or objects flit before my view. When I see them I know I am about to lose consciousness. If they are absent and refuse to come, it means a sleepless night

He worked with Thomas Edison who soon became a rival because Edison advocated inferior DC power transmission system.

Sir Nicola Tesla proved that a magnetic field could be made to rotate if two coils at right angles are supplied with

AC current 90degree out of phase, made possible the invention of the AC induction motor.

Electromotive force is a sort of electric pressure as we see the pressure in water systems, and it is measured in volt, it is called voltage or potential difference, which is present between two points positive and negative.

When a voltage is applied to a conductor or an appliance, current flows through it, the appliance or a conductor resists the flow of current; hence it is known as resistance which is measured in ohms and current in amperes.

Birth of Electricity from visions, philosophy to practicality

Sir Tesla's visions were same like thought process or imaginations of Sir Einstein which is now taking innovators to free energy concepts.

Electricity is researched and generated by means of a machine. Generator is a machine, or a diesel engine or steam engine, which converts mechanical energy into electrical energy, in conjunction with Alternator.

Diesel engine is a prime mover; alternator coupled with the diesel engine. The principle is whenever the conductor loop or a coil is moved through a magnetic field so as to cut the magnetic lines of forces, as the conductor rotates the electromotive force or voltage is induced in the conductor.

The electromotive force changes in magnitude and polarity as the coil is rotated, if there is a single loop the voltage produced will be very small, that is why in an Alternator or AC generator many coils are wounded on an Armature, the AC voltage thus induced in the coils is connected to a slip rings, from which the voltage is received through carbon brushes.

Tesla's philosophy was same as Master Bruce Lee under the roof of one god all human beings are same.

In his own words...

I have expressed myself in this regard fourteen years ago, when a combination of a few leading governments, a sort of Holy alliance, was advocated by the late Andrew Carnegie, who may be fairly considered as the father of this idea, having given to it more publicity and impetus than anybody else prior to the efforts of the President.

While it cannot be denied that such aspects might be of material advantage to some

Less fortunate peoples, it cannot attain the chief objective sought. Peace can only

Come as a natural consequence of universal enlightenment and merging of races,

And we are still far from this blissful realization, because few indeed, will admit the

Reality – that God made man in His image – in which case all earth men are alike.

There is in fact but one race, of many colors. Christ is but one person, yet he is of

All people, so why do some people think themselves better than some other people? It was so nice of Sir Tesla to probe in such a way so we are force to think. Same thought was presented by Swami Vivekananda at Chicago with a title why people disagree he gave an example of frog of the well stating we are all frogs of our little well and appreciated America for their work on world religious parliament for global brotherhood. It takes time to activate because of minds of people working differently

This is when the problem starts and the harmony is lost. It is done directly or indirectly by mind. Mind is coming from a very pure astral world of which our universe is like a tiny hair follicle. That is the reason it is not satisfied with anything of this world. With the dissatisfaction of his mind my father asked me a question. I am quoting it here for knowing the real love. I believe this is the only way for new global government to come and remove the barriers of confining countries, to remove the genocidal and environment repercussions. But human beings should not be made machines or micro chips

This was Sir Tesla's awareness to the people. My father asked me a question which throws light on the similar awareness.

Shekhar Tell me about reincarnation and your masters way about renunciation. When I had told you about mother's love you told me about unearthly love, explain me about this love?

Pitaji

Here is the dialogue from Hindu scriptures to explain reincarnation. Between king and the learnt man Bhanto Nagasena, does rebirth takes place without anything transmigrating or passing over?

Yes your majesty, a man were light a light from another light, pray, would the one light have passed over (transmigrated) to the other light?

No, verily Bhanto

In exactly the same way, Your Majesty does rebirth takes place without anything transmigrating

Give another illustration

Do you remember, Your Majesty, having learnt, when you were a boy, some verse or other from your professor of poetry?

Yes Bhanto

Pray, Your Majesty. Did the verse pass over (transmigrate) to you from your teacher?

No, verily, Bhanto

In exactly the same way, your Majesty does rebirth take place without anything transmigrating.

Bhanto Nagasena, said the king, what it is that is born into the next existence.

Your Majesty, said the elder. It is name and form that is born into the new existence.

My master never advocates the path of renunciation he says you live in the world, perform all the w worldly duties yet try to remain detached.

Within these karmic entanglements we have to find our way for knowing who we are.

My Guru says Mans life does not commence in the womb and never ends in the grave; it is the eternal quest or search for knowing the reality.

In one of those struggling years, I had no job whatever was left. I utilized in two younger sisters classes. Also I utilized so much of my time to leave and take them.

This was due to karmic attachments. I had to perform. After this I had no money.

But my search for knowing myself was going on. The first lesson of yoga I had to give at lokhandwalla complex.

I asked panwalla. The man who sells Leaves used for chewing , how far is this address .Standing at Andheri station , he understood my position and wanted to give money to me which I lovingly refused .

I went walking till that address gave first lesson got money and started my days.

That time I remembered kuntiji's pandawas mother's statement that she wanted poverty because she will remember god, as god was there with her with all the problems she was facing

Many times I had not talked with you to make you understand this and then I wanted you to select wife for me and daughter in law for you so it gives peace to you.

I had to get this done through same sisters who had poisoned our relationship.

Once when I was sailing on a ship named maharishi dayanand. There was a senior fitter he used to talk with his past births son.

At tea time he asked if I believed in life after death. I said yes more than you do also I believe in tramp souls or popularly known as ghosts

He told me your people from hills are doing puja. I told him that usually happens at hills he said no but they are doing it oppositely or in a wrong way. He said that his son says they had been doing this from the time your parents had moved to Mumbai.

I phoned at Mumbai and confirmed that yes they had phoned about that puja. This reminded me of Kashmiri girl who use to fall down standing straight as if she was empowered with some soul or controlled by some tramp souls .I had a strong feeling that this is due to the karmic entanglement of the past births. This kashmiri girl was cured later by one man who was curing peoples from such problems and this girl's schooling also had become better by that, but the man who cured her became sick and eventually died with his neck turning to one side .I believed that, that man had to take the karmas of the kashmiri girl. Once at tea time on ship this senior fitter who talks with his past births son told me .Sir believe it or not your mother is feeding you and your father the tantric things which your sisters are getting. I got annoyed with him when he included my mother. But he said your sisters are giving to your mother by saying it's for the betterment of both of them. I was reluctant to fully believe these because I was following the path of yoga. I knew that these things exist, but I had strong faith on right path and action but I failed to understand the food which I had taken will definitely have certain chemical effects. I was facing many complications which I could fight with the help of yoga. But in spite of all this I was not against them believing this as karmic conflict .I wanted to silently withdrew and within these problems I kept on my search for absolute truth.

Man's life is eternal quest or search for knowing the reality, unless he meets the perfect master who is lord himself, he only can take him to lord. Because that is the only way to meet the lord as it is made by lord himself.

My master says resolving the conflict should be done to the utmost level. We should do our best to avert any bad happening in our families and should take care of our elders in every way.

In large Indian homes there is often a gate house- an independent accommodation built into the main courtyard entrance. Several generations would live together, & as the younger generation would take over the responsibilities of the family, the older generation would shift from the main house to the gate house. So one very cold winter's day, the grandfather of the family asked his grandson to fetch him a blanket. The young boy went to his father. "It's very cold "Said the boy to his father. "And grandfather finds it chilly in the gatehouse. He wants a blanket to keep himself warm."

His father replied, "There is an old blanket in the stable, which the horses use. You can give it to him." The son fetched the blanket cut it into two pieces, gave half to the grandfather & brought the other half back. "There was no need to cut it in two," said his father. "I told you simply to give him the blanket "Well, "said his son. "This half I have kept for you, for when you get old."

My guru clearly states by this anecdote. It is our responsibility to look after our elders, & we should take care of them in every way."

As we have talked earlier about parushrama as pitrabhakta devotee of father , you know I am also a pitrabhakta, but you believed mother as a greatest relation, both are valuable as karmic relationship but father is believed as the protector send by lord for us also I believed ultimate lord is father.

Pitaji if you read the book or works of Param sant Kabir "Anurag sagar" or "Ocean of love" Brahma, Vishnu, Mahesh, sons of divine mother were very eager to meet their father Niranjan.

He had told divine mother no one can meet me. Brahma went for penance to meet his father but he couldn't meet. He lied to his mother about his meeting with his father, when divine mother meditated lord Niranjan refused about their meeting so Brahma was cursed by divine mother. Due to which he is not worshipped; Vishnu and Mahesh told the truth that they couldn't meet their father so they were blessed by divine mother.

This Anurag sagar means ocean of love is overflow of love from perfect Master Lord Kabir to his dear disciple Dharamdas.

These incidents given in Anurag sagar were also narrated by Shri Hajari Prasad Dwivedi ji in his book Kabir

He writes that Sat purush Real God gave permission to his son Kal Niranjan to create shrusti or cosmos or universes. He calls him Dhurt Niranjan cunning one who did it wrongly. He was told to request and take the required material from "Kurmji" the second shabda or son of Satpurush. Kal Niranjan used his might and attacked Kurmji, after destruction of kurmji's body the material came out was for creation

What I read in school about Lord Kabir and kept in my heart. About which I discussed with you Pitaji, but you didn't understood. He described Guru in such a way only Param pita paramatma can be Guru. That is highest fatherly soul which is Real God as described by Hajari Prasad dwivedi. It is the glory of the Guru "Sab sansar kagaj karu. Lekhani sab Vanrai, Sab samudra syahi karu, Guru gun likha na jaya".

If I make whole world as paper, make pen utilizing whole of the forest, Make all seas as ink .I cannot write the quality of Guru. Lord Kabir himself was such a Guru and Dharamdas his chosen disciple.

Kabir for the normal people is an ordinary poet. But one who treads the path of knowing he knows if he never stops with simplicity of heart and soul. Kabir as I read in school was found near the Lahartara Lake to Neeru and Neema named couple. They were Julaha's considered as lower caste. They were earning their livelihood by weaving and making cloth at Banaras. At Banaras Ramanandji was a learned scholar, he was initiating by mantra only to Brahmins. Kabir made a hollow at his pathway to Holy River and slept there hidingly. Ramananda stepped at Kabir's body and uttered Ram, Ram, Ram. I got the initiation said Kabir and ran away; he started telling everyone Ramanandji initiated me. All Brahmins went to Ramanandji how can you initiate a low caste to which Ramanandji said I haven't initiated him. Finally they assembled at court. Kabir was called Ramanandji asked him did I gave you mantra. Kabir said yes Guruji thrice he asked and got angry at a lie. Took out his {khadau} sandal and hit at Kabir's head. Kabir said sir now you have given your sandal of lotus feet's also. Ramanandji was shocked and understood the deservingness of Kabir and said yes I have given you. From then they were to gather. Once while performing worship to the statue of lord. Kabir was sitting outside with a curtain between them .Ramanandji used to put curtain between them. Ramanandji was putting necklace on the statue which he was finding difficult. Kabir said while facing his back to him Guruji open the necklace and perform, to which Ramanandji was shocked and said remove the curtain between us.

Pitaji Mother's love seemingly great it has higher value it has sacrificial feeling for the child .But it is not the real love. Paramhansa Yoganandaji said women are dominated by feelings men's by reason .Both can grow by inculcating the quality of the other. Love as it was between you and me in search of infinite. But Love, Love is beyond mind and matter and grows in the simplicity and purity of the heart.

That Love takes you to Sat guru who is lord himself. Lord comes in the form of Sat guru. Love should be like a moth going near the light.

Kabir explains in love to Dhani Dharamdas O Dharamdas, understanding the reality. I am telling you about love. Those who meditate on Naam given by Perfect Master in such a way that they forget everyone including their family, who do not have attachment of son and wife , and who understands this life as dream , are real lovers.

Brother in this world life is very short, and the world doesn't help at its end. In this world women is loved the most not even parents are loved so much. But the woman for whom one lays down his life doesn't help at the time of death. She weeps for her own self and at once goes to her parents place. Son kinsfolk and wealth are dreams, so my advice to you is to achieve Naam. Nothing goes with us in the end not even the body we love so much.

Kabir in the form of sat guru came in this physical material world at the direction of Satpurush

He was the third kala out of sixteen kalas of shabda Materialized from Satpurush or a third son of Satpurush.

In the beginning Satpurush was in latent form as given in Anurag sagar. He had a desire and created souls. From his first shabda worlds and oceans were created, in which he dwelt. From his second shabda kurmaji was created. Kurmji caught Satpurush's feet and said he want to stay near him. Satpurush gave him the lokas to stay. From the third shabda, a son named Gyan was born also named as jogjit or Kabir. The fifth was Kal Niranjan. The creator of these material world's Sixth son was Sahaj.

The souls living in the meditation of Satpurush were very happy enjoying the nectar. In this way sixteen sons were born.

Beauty of these lokas or worlds cannot be described in languages. Because it is beyond conditioned world of mind and matter the light of these world is not by sun or moon. To describe the beauty and love I am quoting the incidence from ruhani diary of Radhasoami.

The Lord perfect Master
    Shimla 31-10-1944

In bhajan I made some progress. I think. I had told you that I had penetrated the astral world and I had met your Radiant form. From there , after going through many beautiful places .I found the bell sound deepening into vast peals of bells and my vision growing clearer and brighter . I beheld such a brilliance as I could never have thought possible with this earthly mind. The inhabitants were luminous and bright and the dwelling of a design and grace that this earth will never know. I met and converse with the lord of that region. It was scarcely conversation as we know it. Words being almost unnecessary, a great deal by facial expression and gesture and a certain amount by pure perception from there I went into the Region of sunrise and sound deepened to a very deep resounding vibration. It required much concentration to pass through this stage. My artistic tendencies had to be purified. I am very fond of drawing and painting and the colors and forms and views were of a surpassing loveliness which held me down a long time.

I think, I must have explored this entire plane. Finally, I left it and lately more so at the Dera elsewhere. I have felt that vision i.e. earthly vision as such has been shed, feeling as we know it here has gone, and hearing is different. Contact with other souls [who are very bright indeed and very much a part of one, oneself] is by direct perception and above all I feel one with everything I meet [even here when I return to this body] out with you my beloved father, indeed am beginning to realize the meaning of that unearthly love which I have sought for so long. Very humbly I lay this little love at your feet. Several times I have come to the great darkness dear Master but I am still a coward. Please help me with strength. My concentration needs to be collected together very much more than I am doing here and I am weak as water. I have always been afraid of the dark and it will require some doing.

Pitaji it was this unearthly Love I was talking about, when I was under training with naval shipyard. Parents love in greatest in this earth. But when you start understanding logically with intuitive awareness. As I told you about our arti or prayer we were singing in praise of Lord, "Aum jai Jagdish hare" it says " jo dhyayee phal payee dukh bin se man ka " one who meditates, he gets the sukha, contentment. That is of mind. Another stanza is "Para Brahma parameshwar tum sabke swami "Beyond Brahma [this Brahm is not Brahma Vishnu and Mahesh] this is their father Kal Niranjan. Beyond Brahm is the region of real Lord parameshwar, The sound coming from there is swami. When I came from naval dock and sat for meditation I understood the meaning of the aarti or prayer. Sat guru is an abode of happiness. Real Love springs from him. Having Darshan of Satpurush, the jivas were enjoying for many kalpas. One day Dharamdas or Kal Niranjan started meditating while standing on one leg for seventy kalpas or yugas. Satpurush was pleased he sent his son Sahaj to know why Dharam is doing intense tapas. Dharam said he wanted his own creation.

After destruction of Kurmji as given earlier .Satpurush had sent Adhya Shakti or female who had eight parts in her body .Kal Niranjan was staying at Mansarover at that time. When he saw the female and her beauty he swallowed the women. This woman was sent with the seed of creation. Knowing this character of Kal Satpurush cursed him. He will devour one lakh jivas daily and one and one quarter lakhs will be produced.

Sat purush thought of destroying Kal . But then he had to destroy the whole creation so he didn't. But he gave orders to jogjit or Kabir to expel him from Mansarover and he will never be allowed to enter pure regions.

When Kabir reached Kal was showing his powers. But with the power and order of Satpurush .Kabir ji slapped at the forehead of Kal and told the woman to come out from his stomach by taking the name of Satpurush.

Shekhar when you didn't accept at the time of your schooling about earth stands on cow and cow on sheshnaag . How can you accept this story?

Pitaji you are right I don't accept this story either. But we can analyze it hermeneutically that is in the way of understanding. As elders said earth is standing on cow's head. But Upanishads say's earth is based on water and water on fire and fire on air and air on space and space on atman soul seems logical. Pitaji these religious texts are made by hiding the facts so we have to undo them or decode them either by study based on experience and logic or by meditation. As a doctor from Pune wrote to Indira Gandhi about geographical conditions conducive for living at mars .He concluded this by meditation. Paramhansa Yoganandaji made a statement at United States that one gram of flesh is enough to light a Chicago city. Scientist said yes Swamiji is right after experimenting at lab with little flesh they could make heap of flesh, this is experimenting way of modern science. But there was a higher way of going beyond mind and matter .There existed a higher race in Golden era of Satyuga where men's and women's used to live for thousands of years. They had enough time to concentrate on every subject of life to its tiniest detail. Leaving the body was like transcending at will to them. They were aware of higher states of being from land of death. In this Kaliyuga mind is at lesser state and life time span is small. But still awareness has started increasing when thinkers had started coming out of conditioning of religious dogmas. As per Yukteshwar giri the guru of Paramhansa Yoganandaji when the sun in its course of revolution begins to advance towards the place nearest to the grand center called Vishnunabhi. Which is the seat of creative power Brahma, the universal magnetism Dharma the mental virtue, begins to develop this growth is completed in the period of 12000 years. From AD 499 onwards, the sun begins to advance towards the grand center and the intellect of man started gradually to develop .During the 1100 years of the ascending kaliyuga ,which brings us to AD1599 the human intellect was so dense that it could not comprehend the electricity's sushmabhuta the fine matters of creation. About AD1600 William Gilbert discovered magnetic forces and observed the presence of electricity in all material substances, Newton discovered the laws of gravitation, Stephen Gray discovered the action of electricity on the human body. Thereafter development went on towards understanding of the electricity's and their attributes.

As it was said in Bible sun moved around the earth. Later on with study, revolutionary philosophers Copernicus, Galileo came up with that Sun is stationary earth moves around.

Same way meditation is the platform where we can testify the truths in the scriptures. Many of them can be understood logically.

These stories may be fabricated out of facts which we have to decode. Like when we close our eyes. We see a sky. What is that space? Is that space same when we click the mouse of a computer there also we see space. And out in the world there is space. Revolutionary theory of Sir Einstein is based on space and time and their being relative. Before him Sir Isaac Newton saw them as abstract. When Swami Vivekananda went at west he said space and time is nothing but barrier to see the absolute.

Lord Kabir explains in one of his shabda . "meri najar main moti aaya hain, hai til ke til ke til bhitar verle sadhu payahain" I have seen a pearl , inside the dot , there is another dot , and inside the dot there is another dot, rare sadhu knows it. Understanding comes from within. The Kal and the female Adhya created the world, which is the shadow of the shadow of the shadow of the real world.

Woman's nature is such, she never gave knowledge of reality to her three sons and eventually kal eats all of them. Sant Kabir explains the nature of the women.

The Sat guru said;

O Dharamdas, listen to the attributes of women;

I'll make you understand.

When there is a girl in the family. She is brought up with many conveniences.

Her food, clothing and bedding are provided. But everyone regards her as an outsider.

Lovingly performing the ceremonies, she is made to depart with her husband.

When the daughter goes to her husband's home, she is dyed in the colors of her husband.

She forgets her mother and father; Dharamdas, This is the quality of women.

That is why Adhya became an alien, and she, the Bhavani became a part of Kal . That is why, she didn't' manifest Satpurush and showed the form of Kal to Vishnu.

When Lord Kabir came at Satyuga, His name was Sukrit. He gave Naam to King Dhondnal and khemsari they understood the illusions and were liberated. Lord Kabir is Anami about which he described in his shabda. He came in every yuga by the orders of Satpurush , to remove illusions created by Kal the negative power created by Satpurush . At treta yuga he gave initiation to Vichatra Bhat and wife of Ravan named Mandodari At Dwaparyuga he initiated Queen Indumati. And at kaliyuga he came as Kabir. At Dwaparyuga he gave Naam to saint named Supach sudarshan. Whole heartedly he did the devotion of the Sat guru, leaving all deceptions and cleverness. His father and mother became very happy and in their hearts they had great love for him. When Supach went at Satlok. He requested Lord Kabir to liberate his parents as they were suffering in the Land of Kal. He requested by saying "I explained to my father and mother in many ways but thinking me as child they never believe me. But they didn't stop me from devotion. They were very happy with me.

They were always pleased with me. That is why. O Lord. I make this request to you, bring them after making them firm in the Sat Shabda, and cutting their attachments, liberate their souls,

By the Glory of Sant Supach they were born into a Brahmin family named kulpati and Maheshwari.

Kabir materialized himself in the form of a child to Maheshwari. Both husband and wife were serving happily. Kabir explained them shabda but they didn't believe in the child so Kabir disappeared.

Their second birth was Chandansahu and Uda. Again Lord gave her the Darshan as child. She brought the child home but her husband didn't agree to keep the child. They got the third birth as Neeru and Nima where they found Kabir sahib at Lahartara Lake.

Kabir ji spoke to Nima, hearing Kabir ji she feared of Neeru. She took the child home. But still later on they forgot the shabda.

Their fourth birth was at Mathura and they went to Satlok.

Way towards personal safety and social responsibility

Before controlling these finest particles about which material science has worked so far. We have to control and know or are aware of our mind; one's mind is under control we can control the life force or prana. This is about Vajramukti a combination of yoga meditation and movement art. It is imperative to practice for stress management. Real pssr personal safety and social responsibility since I had been working on naval ships and shipyard I had experienced that in reality pssr never happens. .Vajramukti is a combination of yoga and martial arts. It was a way of self discovery for me and my students. We were practicing meditation by not touching psycho physiological centers or chakras. Like for example in kriya yoga ujjayi pranayama is done while concentrating on the chakras but we do it without concentrating on chakras we insist more on prolonging the inhalation and exhalation. Ujjayi pranayama is done by partially contracting glottis and epiglottis as we exhale with producing the sound in martial arts. This is done slowly so that for some time you don't require breathing which relaxes your heart and in turn affecting other organs of the body. Before doing moving meditation or vajras or katas we were concentrating on medulla oblongata or small brain and contracting different parts of the body visualize that energy is passing from cosmos through back of the head to center between the eyebrows. From this point it is redirected to the part of the body to be reenergized. This I took from kriya yoga about which I am explaining with the help of postures

This is known as jivabandh or tongue lock. You can sit in any posture but back should be straightened. In this posture I am sitting is padmasana also known as lotus pose, you can also sit in sukhasana crossed leg pose done before this where tongue is stretched outside .Ancient text have described it as destroyer of all disease, 72,000 nadis are purified by performing this asana. Pull your tongue fully upward towards the upper palate of your mouth and stretch. Then push it outwards as given in the posture extending the arms and stretching the fingers is optional. You can do this before doing ujjayi pranayama for around thirty times

Pulling in and pushing out of tongue stretches the ligaments supporting the tongue. In kriya yoga it is practiced to fold the tongue upwards and backward towards the hollow portion, to receive the nectar from the higher source. But for this one has to practice extra and under supervision. It has a profound effect on vagus nerve, which is the largest autonomic nerve and it sends fibers to many glands and organs. The autonomic nervous system is divided functionally into two parts, Sympathetic & parasympathetic system. Sympathetic system lies in front of the vertebral column and is associated and connected with the spinal cord by the nerve fibers. The parasympathetic system is divided into two parts, composed of the cranial and autonomic nerves. These sympathetic & parasympathetic cords constitute the autonomic system which supplies nerves to the involuntary organs such as heart lungs kidneys liver etc and controls them. The vagus nerve is the tenth cranial nerve and is associated with the medulla oblongata. By practicing these it is observed that mental or mind related problems are solved.

These two poses given also helps removing wrinkles from the face naturally it makes you aware about the throat. While practicing ujjayi pranayama concentration is to be done at throat

UJJAI PRANAYAMA

Sit in folded leg posture put you awareness at throat. Inhale through throat by slowly contracting the throat area. T he sound produced is AW. Increase the time of inhalation and exhalation. It calms the nervous system, controls the system resulting in mental relaxation can be used for controlling high blood pressure.

Let me explain how to sit in lotus posture. Sit on the buttocks, place the right foot on left thigh and left foot on the right thigh, palms at the junction of your thighs or at your knees. Initially for few days you can practice by keeping right foot on the left thigh and not left foot on the right thigh just fold it beneath the right thigh. Slowly as your body tunes with relaxation, through removing of thoughts you happen to do it. There is no exertion involved in yoga.

You can also practice these in vajrasana. Sit on your buttocks extending legs at front, fold one leg at knee and take below the same thigh, do the same with other leg. In this pose the spine comes in its natural position i.e. in the form of question mark. It controls the vajra nadi, removes excessive gas from the system. It gives extra power to the digestive system. This pose can be practiced also after heavy meal.

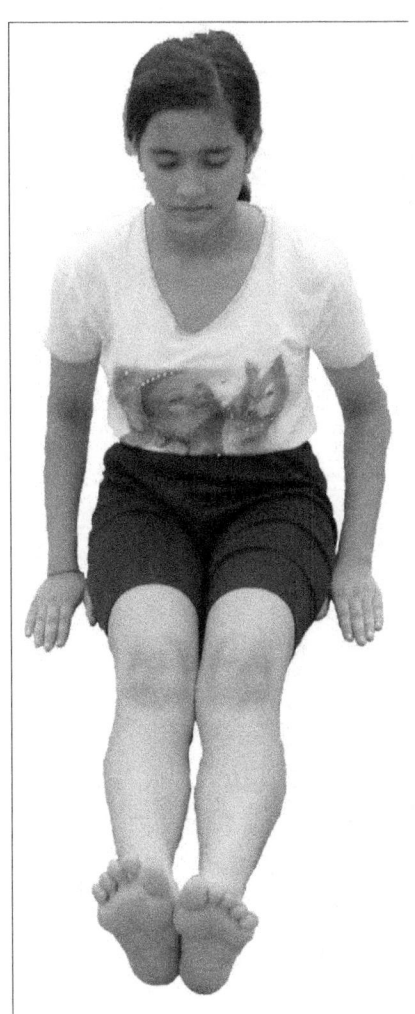

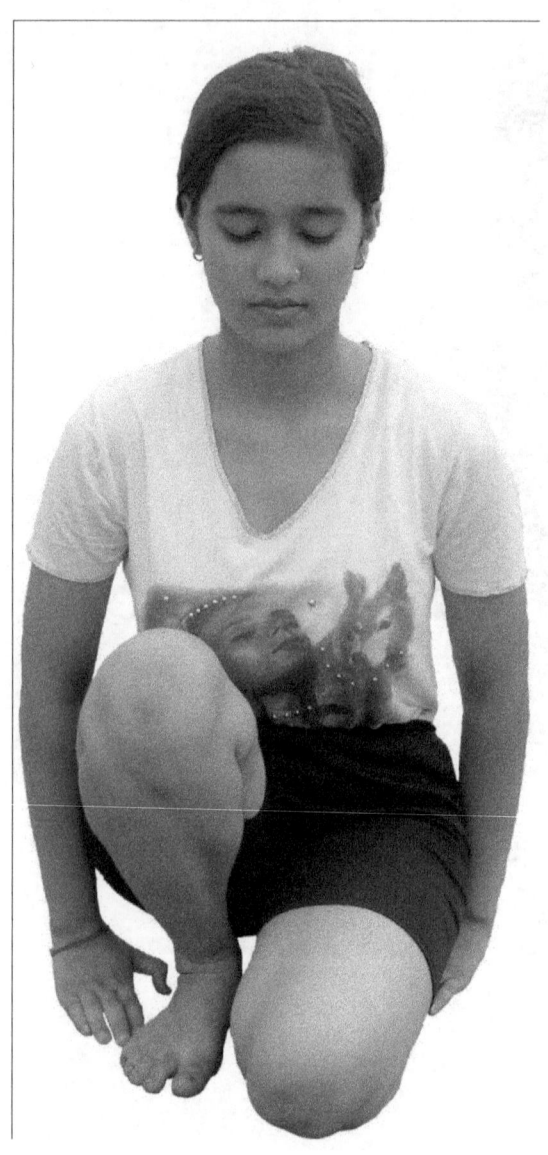

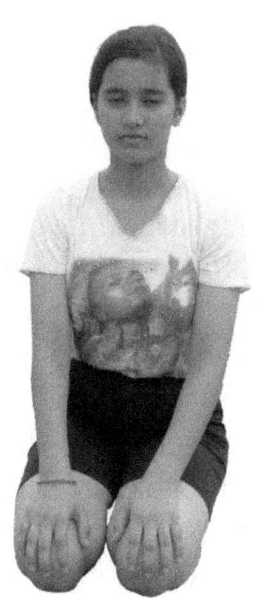

You must learn how to differentially relax your body. I am explaining this with the help of one leg up pose known as utthitekpadasana. Lying on your back lift only one leg, feel that your other body parts are relaxed by slightly bending your other leg, which is not doing the action. After attaining the pose put your mind on breath count the breath. One inhalation and exhalation makes one count

In the poses below forearm and calf muscles of left side are contracted while doing so other muscles remain relaxed, while releasing visualize energy coming from back of head, through center between eyebrows to forearm and calf muscles and recharging the cells. In the second pose the upper arm and thigh muscles are contracted in the same manner.

This is another pose for cell recharging .Fold your knee upward & kick sideways slowly contracting the muscles of the leg while concentrating at eye center or between the eye brows. After contracting release the muscles feel the vibration & energy pulsating as tingling sensations.

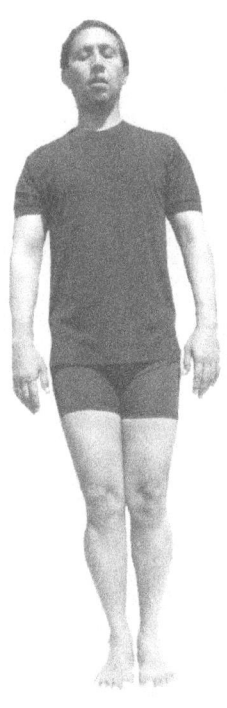

In these two posses first exhale through the mouth then inhale and start contracting the muscles upward from feet's calves' thighs buttocks lower and upper abdomen take the palms at the sides facing upward. Go on contracting shoulders neck and facial muscles till the scalp area keep it for twenty seconds then exhale releasing the muscles in reverse order as in the first pose. Feel that all body parts are recharged.

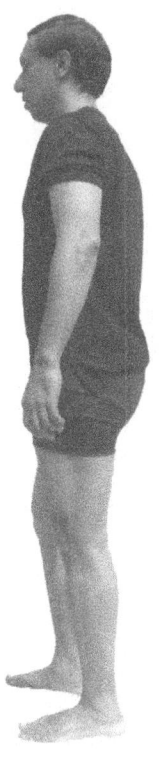

In the above pose put your body weight on left leg concentrating on your left buttock and chest muscles so that you can contract left side buttock and chest muscles. Perform same way on the right side while releasing feel that the energy is coming from back of the head through the place between the eyes and pulsating at the part which you had relaxed.

These two poses are kind of spinal adjustment. First rotate or move the upper portion that is shoulders to one side then move lower portion that is hips to other side. Once you understand this, move both together. This will remove any minor maladjustment in the spine.
Viraabhadrasana is another pose to correct the spinal maladjustment.

VIRABHADRASANA

 Two poses are known as Viraabhadrasana
named after a martial hero virabhadra. These
are also helpful in spinal maladjustment spread
your legs and hands sideways. Palms facing
downwards stretched hands as if they are
pulled. Turn face to one side front foot straight
back foot turned inward bend front knee shin
perpendicular to ground. Stay in pose for short
time or depending upon your enduring capacity.

This is the third pose of moving the series. I had not cropped this pose so it looks different because here you twist your torso or upper body or hips touch your palms together tilt the head this gives the intense stretch to spine and hip muscles it also helps in removing minor tilt of the posture. These Viraabhadrasana and moving meditation can be done after long hours of sitting meditation.

Place body weight on front leg by bending forward for preparing the easy lift of back foot straight at knee

This is for side stretch. Place the palm on the floor while standing in Viraabhadrasana pose slowly lifts the leg straitening the knees

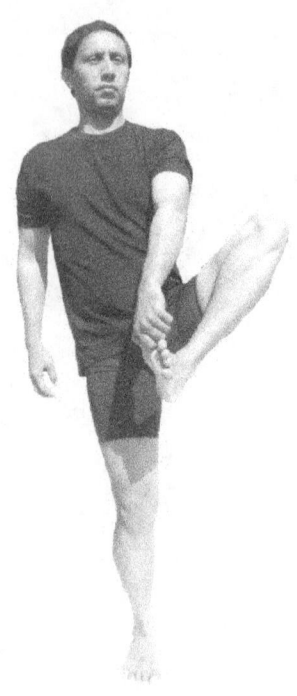

These four poses above are for increasing the sense of balance and stretching. Practice with both the legs. These all balancing stances or what I call is virabhadra are important for moving meditation

Moving meditation should be done after long hours of sitting meditation. I am explaining her how virabhadra are moved and body is carried from one part of space to another. In virabhadra the more weight of the body is on the front leg. This is just the little idea one can shift weight as per his requirement. You can put more weight on back leg and create your own moves. These movements are known as vajras they are nonviolent moves can be used for self defense. Main purpose is in aid to meditation. These are means towards end and not end in itself. These should not become a conditioning

Master Bruce lee said is style important or individual. Both has their value but individual is more important. The system which is moving towards de conditioning is more important.

Here moving meditation is started with vajrasana

Meditate on vajrasana index finger touching the
thumb. Slowly rise while inhaling and attain
Viraabhadrasana on either side turn your wrist
inwards. While doing this feel the sensation as if
energy is coming through medulla oblongata or
small brain and getting distributed through
center between eye brows. Bend your wrists

inward feel the energy flowing as vibrating sensations.

Make an arc with back leg and turn exhaling.

Complete the exhalation by feeling the energy and turning the wrist inwards. Hold the breath for few seconds and come back in normal position

This is just an approach towards feeling of energy. It is not a conditioned way. One can make his own way because every individual is different. I will give a little idea of how to move.

This is when you want to turn on say left side. Put more weight on your right leg take your left leg sideways toe touching to ground. slowly turn to left side.

Same way turn on right side

While practicing these you can initially do inhalation and exhalation alternately. When you feel comfortable you can do two times inhalation and three times exhalation. Then four times exhalation. Idea is to make exhalation double of inhalation or increase the time of exhalation. This can be done by any one old or young. Holding the breath after inhale is dangerous if you do beyond your capacity. Holding the breath after exhale is not that dangerous. Prolonged inhalation and exhalation as in ujjayi pranayama in kriya yoga gives the same effect as given by holding the breath. Further poses I am explaining about breathing process. I had worked these poses with managing director of International casting. His doctor said within few days he will die. But after practicing these vajras his heart became better. I worked with many people and found them beneficial

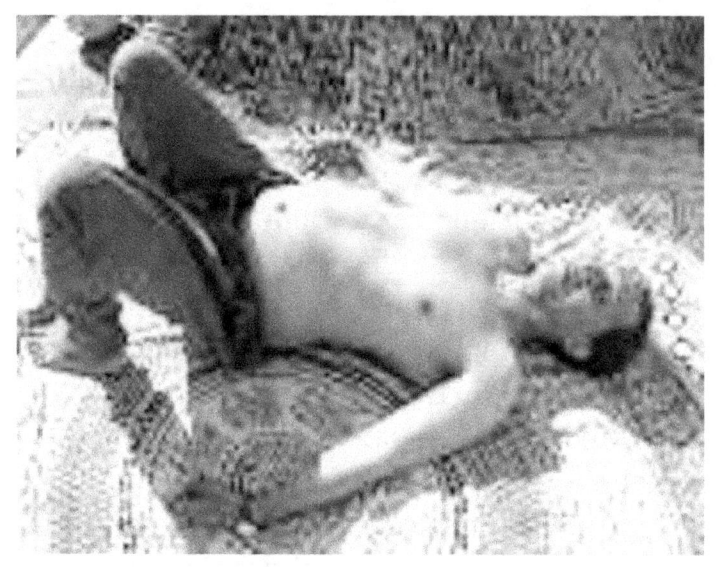

137

These four poses are known as tadagi mudra. Meaning lake like In the first pose inhale fully expanding the stomach .Generally if you are aware when we inhale first our stomach is protruded and then the rib cage and thoracic region finally the shoulders exhaling is in the reverse order. In this pose after inhaling and expanding the stomach exhale and push the stomach outside further and hold your nose now let the stomach drop and go inside on its own. Keep for few seconds holding the breath and then release. This is the preparatory pose for any pose which requires abdominal lift such as Nauli or uddiyan bandh

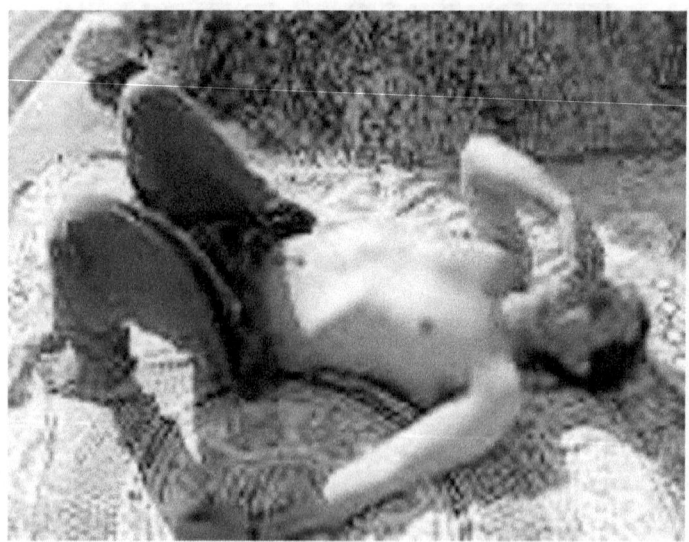

GENERAL AWARENESS OF BREATHING PROCESS

We should be aware of generally how we breath, normally people know that we breathe by expanding chest, as we don't fill in our lungs to its fullest capacity neither we expel the breath to its fullest capacity, the residual remains our body and brain both are starved and if the residual remains it may be dangerous leaving body for other diseases of lungs.

So when we breathe, first the abdominal muscles are expanded then the rib cage, chest and shoulders are lifted slightly and while releasing the reverse is done, by doing this we are using abdomen and thoracic or chest breathing, by doing this lungs get the optimum amount of air removing the residual to its fullest which prevents the lungs from any diseases.

PRANAYAMA

Prana is a vital force which is in the whole of creation, the air we breathe is not exactly the prana. Prana is subtler it is an essence or the vital force by which even the air is lively, yama means to control, so to control prana is pranayama.

Just holding breath may be dangerous, for pranayama one should go through some expert, or do it as easily as possible, we see when a person is engrossed in some work, he generally breathes very slowly, quick and uneven breathing happens at the time of anger, fear or lust. The monkey breathes at the rate of 32times a minute, a man's average breathe is 18 times a minute, the tortoise who lives for 300 years just breathes 4 times a minutes.

Prolong inhalation and exhalation as in bhramari pranayama also gives the same effect, not necessary that you hold the breath for longer time, as in kriya yoga there is no holding of breath but the effect is great, the blood is de carbonized and recharged with oxygen and the atoms of this extra oxygen are transmuted in to life current to rejuvenate the brain and spinal centre's, by stopping the accumulation of venous blood, the yogi is able to lessen or prevent the decay of the tissues.

People with high blood pressure or any other problems should consult the doctor before starting the practices.

Prana is divided into 5 pranas

PRANA: Controls respiratory systems and speech organs, activates inhalation and exhalation.

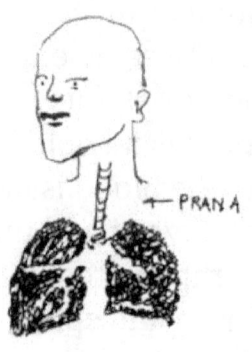

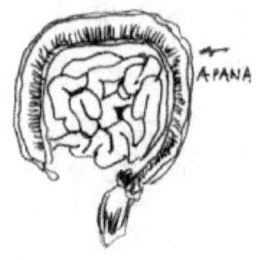

APANA: Located below the naval region and activates expulsions and helps eliminations

SAMANA: Activates and controls the digestive system, it is concerned with the region between the heart and the naval and helps assimilation of the nutrients

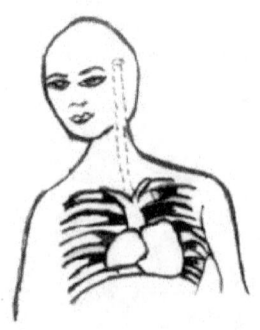

UDANA: Dwells in the thoracic cavity and controls the intake of air and food, helps metabolism.

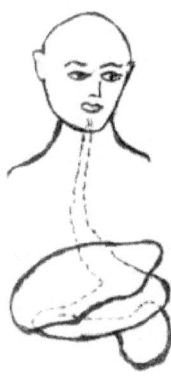

VYANA: It is a vital force, pervading the whole body harmonizes and activates the feet, hands and other body instruments and their associated muscles, ligaments, nerves and joints

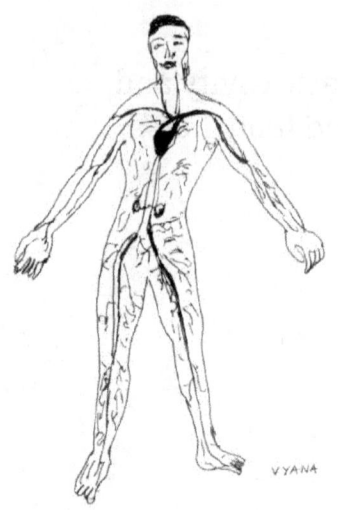

VYANA

UDDIYAN BANDHA

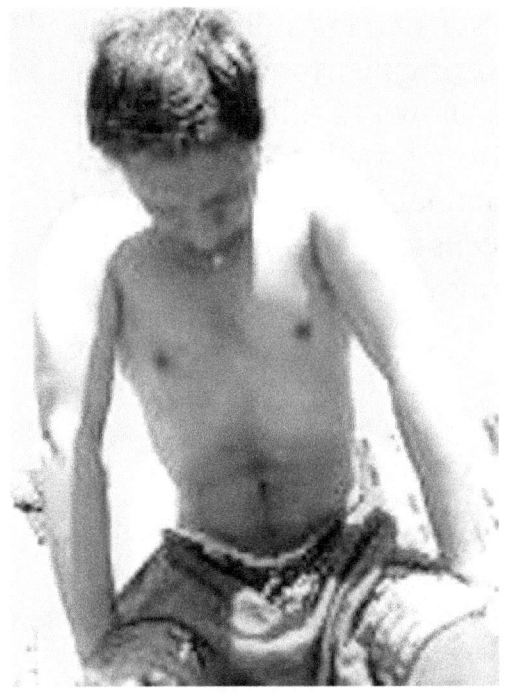

As explained earlier Tadagi mudra or lake like pose helps in performing Uddiyan bandha meaning abdominal lift. Here it is done by bending the knees. Placing the palms at mid of the thighs inhale fully by protruding stomach then exhale by pushing the stomach further as in tadagi mudra. Hold the breath and let the stomach go inside. In this position while holding the breath if you start taking stomach inside and outside or moving the stomach rhythmically it is known as AGNISAR KRIYA which is preparatory pose for Nauli the churning of abdominal muscles. This control wind bile and mucus which results in rejuvenating the system.

NAULI

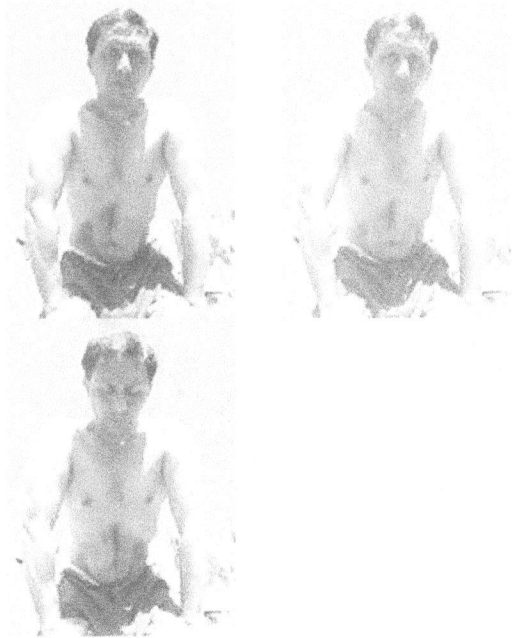

When you will put pressure on one of the palms kept on the mid of the thigh. One side of the intestinal muscles will be affected and protruded. Practice same with the other side so other side of muscles will be protruded. If you put pressure on both the palms only the mid portion will come out. With the help of both the palms moving alternately you can make the muscles move circularly which is known as NAULI.

JANUSHIRASANA

This pose is known as Janusirsasana in hath yoga. In kriya yoga it is performed as Mahamudra. It prolongs the time of cell disintegration thus keeps the person rejuvenating and full of vigor. Bend any one of your knee and place your feet at inner thigh the other foot straight. Exhale take stomach inside this is known as uddiyan bandh and hold the breath .Contract the area between anus and scrotum this is known as moolabandha. Touch your chin towards chest this is known as Jalandharabandha try to hold your extended foot as easy as possible and touch your head to the knee. Slowly release the breath and relax. These bandhas or locks have profound psycho physiological effect in your endocrinal system. They are infinitely better than modern exercises .These was used by Aryans who were believed to come from outer space. They were highly evolved beings. It depends on the reader if he wants he can take or leave it. Because even prolonging life span has no meaning if you don't know thyself. These bandhas or locks can be done in conjunction with mudras for better effects. Science of mudras is an ancient subject put in front of mankind by the rishis for the better and healthy living, according to them the secret of health and rejuvenation lies in hands.

Mudras are performed with the help of fingers, these are condensed in the traditional philosophical dances, also were performed in martial arts and Indian rituals.

Our body is made up of five basic element earth, water, fire, air and space, in order to keep the body healthy any disturbances caused in the body can be cured by performing these mudras and bandhas.

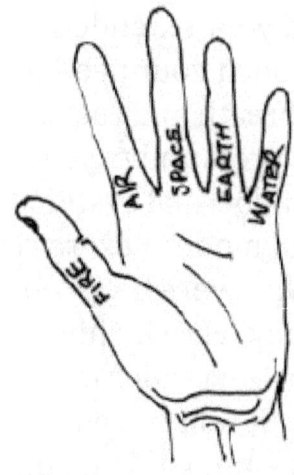

Five fingers of the palm controls five basic elements thumb—fire, index finger—air, middle finger—space, ring finger—earth, small finger—water.

Cow posture, fingers of both the palms should be placed in a way that both the ring finger touches the tip of the last smaller fingers, crossing each other, then middle fingers touching the tips of the index fingers, as shown.

Benefits in memory stimulating and improves the memory power.

Dhyan mudra

Touch the thumb with the index finger, these mudras can be performed in any easy postures.

Benefits: It is the most important mudra used for meditation, helps in increasing brain power, prevents mental disorders, mental tensions.

Prana mudra

Prana

Bend the little and ring fingers so that, their tips touch the tip of thumb as shown.

Benefits for improving eye power increases life force ki or prana

Linga mudra

Entangle the fingers of both the palms keeping either of the thumbs erect.

Benefits in case of cold and bronchial infections

Vayu mudra

Touch the index finger at the base of the thumb that is the part of Venus as shown.

Benefits, it prevents rheumatism and purifies blood.

Sunya mudra

Shunya

Keep the middle finger at the top of the Venus as shown in fig and press lightly with the thumb.

Benefits, it affects the person who is weak in hearing, also beneficial in vertigo.

Varun mudra

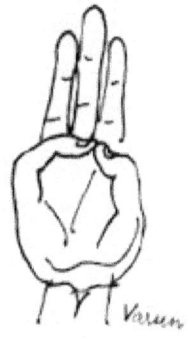

Touch the tips of thumb and little finger together as shown.

Benefits in gastro intestinal problems, helps skin to become smoother.

Prithvi mudra

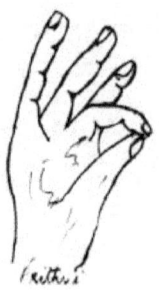

Touch the ring finger to the thumb as shown.

Benefits in maintaining the balance of earth element in our body, strives for eradicating physical weakness.

Sun mudra

Bend ring finger so as on its outer side on second fold, you can press with the thumb as shown.

Benefits in weight reduction and fats in the body

Yoga mudra

Sit in padmasana or vajrasana grasp one wrist behind the back with the other hand, slowly bending the upper body, bring the head to the floor, touch the forehead to the floor, remain in the position, slowly assume the starting pose and relax.

This is preventive for abdominal ailments and vertebrae related problems.

These yogic movement amalgamated with martial arts is good for all human beings specially for the ones who are far away from the society and nature like sea farers

Kaki mudra

I

n this mudra the lips are pursed resembling a
crows beak, can be done in any meditative pose,
this mudra is very good for overall development

of the face, by regular practicing cheeks become red. It is a age defying balm.

Close your both the nostrils with both the thumbs, fingers touching their tips, start inhaling through the lips, open your eyes while inhaling, after inhaling, expand both your cheeks, touch your chin to the chest and hold your breath as per your capacity.

After performing those bring your head to the normal position, slowly open your eyes & release your breath through nose.

Simha mudra

This is a symbol of lion, practicing this mudra reduces the wrinkles from the face, effects are much better if done in conjunction with jivabandh, tongue lock taking the tongue to the upper palate of the mouth and also stretching it outside as shown.

Controls enlargement of tonsils, stabilizes blood pressure.

Sit on vajrasana on the heels, open up the knees resting the body on arms, tilt the head backward, extend the tongue by opening the mouth open, eyes wide gazing at eyebrows exhale producing a h sound.

Vajroli mudra

Practice of Vajroli mudra controls the vajra nadi by which sexual energy is controlled and semen retention power is increased. This posture is sukhasana easy cross leg pose.

Sit in any meditative pose or sukhasana, put your awareness in sex organs for males, contract the penis and pull upward for females contract the clitoris, lower vaginal muscles and urethra and pull upwards. Hold for few seconds and release. Initially one tends to contract the entire muscles anus penis and the place between scrotum and anus. But with practice you can segregate them

Ashwini mudra

This is preventive in the case of piles, sit in any meditative pose or sukhasana put your awareness at the muscles of anus contract the anus and release, do it as easily as possible without straining. This can be done in conjunction with Pawanmuktasana for prevention of piles.

Moolabandha

Is useful in both the purposes to attain a celibacy and to prevent the sexual problems, it takes the sexual energy upward for spiritual development or downward to enhance marital relations.

It controls the life force consequently psychosomatic degeneration, is controlled increases vitality, also increases the sexual retention power, and should not be practice in case of absence of periods for females.

Moolabandha may be practiced at the time of pregnancy for increasing the elasticity of the muscles also after child birth, along with vajroli mudra and aswini mudra for restoring the muscles and controlling the neuro muscular system, it generates excessive sexual energy, prevents hernia, and controls testosterone secretions.

Sit in any meditative pose or sukhasana, put your awareness at the place between the scrotum and anus, contract that place and draw the muscles upward, initially it is difficult to draw only muscles of that area, all three gets contracted but in due course of time it happens.

Before explaining Kapalbhatti let me explain half kapalbhatti or also known as suddhikriya. Close one nostril with your thumb and expel or exhale through other nostril taking the lower abdomen inside. Do not force after doing one stroke wait for the air to come inside naturally then go for other stroke. Can be done ten times through each nostril do three sets or as per your capacity if same is done with both the nostrils it is known as kapalbhatti.

CHECK This I am writing specifically because it applies to all asanas and Pranayama. Also I have seen many people performing incorrect. If you fall short of breath after performing then it is done wrongly.

Sit in any meditative postures or sukhasana, expel your breath forcefully through your nose taking your stomach inwards, do it initially for 50-60 breaths then relax after some time, perform moolabandha and do the same practice. Gradually increase it to thousand exhales.

Benefits: it prepares you for meditation, purifies frontal region of brain as the name suggests kapalbhatti, it shines the forehead, beneficial in the case of cerebral thrombosis.

BHASTRIKA PRANAYAMA

Sit in any meditative posture or sukhasana perform moolabandha. As you perform moolabandha your lower abdomen is locked. Now you can move upper abdomen keeping the body steady, hands rooted or kept on knees as in kapalbhatti breathe rapidly. Inhaling and exhaling rapidly using upper abdomen. Initially do it slowly so that counts can be increased after around fifty or hundred counts. One count is one inhalation and exhalation, exhale fully then inhale through right nostril hold the breath touch the chin to the chest and retain as easily as possible. If you have done moolabandha also while retaining the breath then before exhalation release moolabandha. Exhale through left nostril. If you can not perform this practice the easier way is by inhaling and exhaling by concentrating on chest do not move the shoulder. Hands rooted at knees by this also you will feel you are moving upper abdominal muscles gradually when you understand this you can perform as explained earlier.

Benefits: it removes the impurities from the lungs, benefits in asthma, tuberculosis

BHRAMARI PRANAYAMA

Sit in any meditative pose or sukhasana, spine erect plug both the ears with index fingers. Mouth closed, take breath through the nostrils, hold it for a second and start releasing the breath from nose producing m mm sound while mouth closed slowly release the hands and relax.

Benefits in blood pressure problems, develops the sound quality or the voice, relieves mental tension.

ANULOM VILOM PRANAYAMA

Sit in any meditative pose or sukhasana, spine erect. Closing the left nostril exhale through right nostril and then inhale through the same right nostril. Hold the breath for few seconds then exhale through left nostril hold for few seconds. Further inhale through the same left nostril and exhale through right nostril. This completes one round of anulomvilom. One can increase as per his capacity. Very good for balancing the system and improvement of eyesight.

SURYANAMASKAR

If you have low blood pressure you can start with suryanamaskars. It includes ten steps with successive inhalation and exhalation. By practicing these King of Aundh was full of vigor at the age of 110.Also bhastrika and kapalbhatti pranayama can be added after practicing asanas for low blood pressure. For high blood pressure Pawanmuktasana, ujjayi pranayama and savasana will be helpful

Feet's together with folded hands thank to that one lord. Stretch the hands upward creating a slight arc at back INHALE.

Feet's together with folded hands thank to that one lord. Stretch the hands upward creating a slight arc at back INHALE.

Touch the palms at floor try to keep knees straight EXHALE

Take the left leg back keeping right feet planted,
right leg bend at knee INHALE

Take the right leg also back; both the palms rooted to the ground hold the breath for few seconds.

Bend the elbows drop the forehead to the
ground touching the ground, chest knees and
toes of the feet's touching the ground EXHALE

Drop the
buttocks release feet's at floor lift the shoulders
tilt the head backward hands rooted to the floor
INHALE

Lift the
buttocks up palms and feet's rooted to the
ground EXHALE

Take the left
leg between the palms right leg outstretched
INHALE

Bring both
the feet's together palms rooted to ground
EXHALE

Stretch the hands upward as in the pose one INHALE

This is very important for females, it will make their breasts firm, remove extra fats from the buttocks, will make their legs shapely as in other exercises they may develop their calf muscles as men's, this will not happen while practicing suryanamaskars, salutations to sun flexibility and strength both will be achieved.

It removes acne and clears the complexion because the impurities and toxins are thrown away from the body, the body of old king of aundh at the age of 110 was like a youth, by regular performing of suryanamaskars.

In today's modern, fast life, we don't have time for deep breathing exercise, generally, our mind is engaged in some thought, consequently the breathing becomes short, for females the situation is even worst, due to the better psychological conditioning done by media, they are attracted towards faster exercising clubs they join fast and leave fast and if continued mental stress develops.

Surya namaskar is such ancient yogic movement, which enables you to breathe deeply and it powerfully influences your endocrinal system which controls the body functioning and makes it better.

Yoga is the base of martial arts as the meaning of the name is to join the soul with supreme soul. All artists fall under one roof when logically they start understanding what lies in their innermost selves. Martial artist Gichin fuakoshi stated if you jump and kick throughout your life you become a puppet unless you realize what lies in your innermost self.

Artist can be from any field, like modern artist Frirzof Capra writes in his book Tao of physics.

Five years ago, I had a beautiful experience which set me on a road that has led to the writing of this book .I was sitting by the ocean one late summer afternoon, watching the waves rolling in and feeling the rhythm of my breathing, when I suddenly became aware of my whole environment as being engaged in a gigantic cosmic dance. Being a physicist, I knew that the sand, rocks, water and air around me were made of vibrating molecules and atoms, and that these consisted of particles which inter acted with one another by creating and destroying other particles. I knew also that the Earth's atmosphere was continually bombarded by showers of cosmic rays, particles of high energy undergoing multiple collisions as they penetrated the air. All this was familiar to me from my research in high energy physics. But until that moment I had only experienced it through graphs , diagrams and mathematical theories.

As I sat on the beach my former experiences came to life I "saw' cascades of energy coming down from outer space, in which particles were created and destroyed in rhythmic pulses: I " saw" the atoms of the elements and those of my body participating in this cosmic dance of energy: I felt its rhythm and I heard its sound and at that moment I knew that this was the Dance of Shiva , the lord of Dancers worshipped by the Hindus.

FROM SUBATOMIC PARTICLES JOURNEY TOWARDS UNKNOWN

Prana is as visible as the electricity is, we can feel it in the air but is not the air, it is by which the air is alive. By controlling the mind alone we can get control on the physical .Manifestation of the particles, all the natural forces. We can materialize and de materializes our bodies as yogis do. Electricity is like a Mumbai city meaning it is vast or in other words its attributes had affected profoundly to the human society, it is subtlest of all the sciences. The study of electricity or electron has made physics more vital. It happened to me the guiding post for the path my search for knowing the ultimate a means towards Ultimate.

It is a search for materialistic in its own way towards infinite or could be said a search for the primary substance out of which all world are formed. Electricity and rays would be the future medicine of the world. A dreaming person does not know that his dream of a city containing buildings, streets and lights are made up of different vibrations of his imaginations, frozen into thought forms. Science proves that all matter is electromagnet waves, and these waves are frozen light.
Metaphysicians say that these waves are frozen light in frozen God- vibration. But we as small light have to experience it on the platform of meditation for our own self content.

I won't call the way of electricity a path of heart as teachings of Don Juan says because heart centre in our body, is a place where mind dwells. A lesser degree of mind, whereas when we go for search inside of our being, from the lowest chakra to the highest in the body that is thousand petals lotus. The mind goes on refining ,and at this state it is refined to a very high extend , yet one has to go much higher, then how can I conclude for the path, at heart chakra, but still the idea is good and helpful for growing as he puts

"Look at every path closely and deliberately. Try it as many times as you think necessary.

Then ask yourself, and yourself alone, one question ... Does this path have a heart? If it does, the path is good; If it doesn't it is of no use". By Carlos Castaneda, from the teachings of Don Juan.

The purpose of this book is to present my awareness to my people from various religions.

A search of mind which started at the time of my training with Indian naval ship yard

Yoga and Martial arts were also going parallel with this.

I found that none of these, physics to the subtlest levels and all religions can take you to your creator. All philosophers Socrates Pythagoras and others said know thyself man

Way of electricity was a search towards the path, the exciting search through nature, which gives knowledge about physical universe and its underlying principles, make you aware to probe into unknown which leads you to search for the path, any path you follow will be the path of heart at level of your awareness, you shouldn't be conditioned even by the path. The path has to be tested and also the master so far as we can satisfy our intellect.

At the time of my training in Indian Naval shipyard I and few of my friends sitting for discussions, a sort of class I would call Vajramukti interfaith dialogue class, Vajramukti meaning a way which takes you from action to libration.

Sitting under the trees in open, there is small hut where eating stuff and tea is being served, the owner and maker is jain a follower of Mahavirjain he makes nice tea we take vadas along with the tea. Vadas round shaped eating stuff made of potatoes vegetables and dough served with powdered garlic chutney.

Unni a friend of mine is helping himself with vada a junior Mathew is sitting and listening to my talks about Jesus Christ I explained him about my talks with martin

Mathew is a nice man he wants to change his profession to something where he gets more time to himself, for knowing himself or meditating.

Unni helps himself to some more vadas he had to say something, I understand by his gesture, talk is about people like labors who are at time more intelligent than the trained and educated one, but they remain labors only, why is it so. One of them was dear Kalidas a follower of Radhasoami philosophy. I used to meet him regularly for philosophical discussions. He told me try to catch sound currents and you are into process of meditation. Martial artist Bruce Lee was using it for getting video audio and tactical cues while practicing at arena.

I tell Unni about the philosophy of karma, since he had studied in convent school his process of thinking is more like them, I tell him about my past birth that I was Chinese all of them looked at me with surprise in their eyes. Unni is carefully silent since he is more diplomatic. Vimal understands and the silence pervades. Suddenly we are all not there at the present moment. We are all alone in our private universes. I understood and I wanted them to be in their own worlds. This was what I used to call Vajramukti interfaith dialogue. So that next time they will come up with their own ideas that will be real Vajramukti freedom through action. I bid a good bye to them because I wanted to get away from this silence which speaks from inside and go to utter silence, where I see the inner world through meditation.

To talk about Kalidas and other labors and stupidity of the system is an attack only to the fragments or the branches of the tree, like watering these branches, no nourishment or no change will be possible in this way. The main thing is to water the roots or to remove the cause of ignorance, or increase the awareness of people, the true way would be to increase or to help to increase the awareness of the people, the revolution destroys a systematic government, but the awareness will help or enhance the thought process of the people hence a slow change will be set forth into practice which will be of no loss to individual consciousness and collective consciousness, that is family and society and Nation as a whole. Considering individual consciousness I am coating Shri Rabindranath Tagore. He had put an impression on collective consciousness when he refused to accept doctorate from Government stating the mental slavery of the government towards Nobel Prize.

He said you are conferring this doctorate to Nobel Prize, his book geetanjali was lying in country much before than when he got Nobel Prize may be this way of thinking has developed by mistakes he made writing in praise of British officials. Like Pratap Antony writing on his blog about adhar card and making people aware of their rights.

My martial arts teachers were two brothers Sensei Oliver and Sensei Leslie Fernandez, Leslie was more softer and interested in philosophy, Leslie asked me a question, can anyone defend himself by bullets he would not have expected this answer, if someone moves subtler to subtler strata of his being , yes he can protect himself from bullet attack, but at that level may be he won't need it any more, there was something in me which was always aware about the super natural powers, but I had lost it now, somewhere in distant past, I had some power which I tried to put into experience .

Today it is Tuesday we are heading to Brazil the thoughts are coming of past births, at times I could feel that, king Harshvardhan gives blankets and many things to the poor people, and I am the Chinese traveler attracted by the Indian life style and culture, which was the strangest of the world as writes Marco polo in his works that those Indian Brahmins have sex only with their wives.

After 10 years I visited France again, one thing was there in my inner being or in other words my inner being was saying there was something which was missing there, in spite of all the material richness, and I knew it was soul. I felt there is no soul or lacking of light of soul there and then I realized yes the light of the soul was at lesser degree there, while at INDIA because of the Masters living here in form and yet they have gone to the highest level beyond, I have felt the soul with more light and more feeling is here.

When king Visayana was ruling Philippines at the time of Maurya dynasty in India, at that time Chinese travelers Hyun sang fayan had visited India, Visayana took with him the astronomical calendar and system of measuring which was very minute at that time after Muslim invaders came all that was finished.

Same way in today's world system of weights and Measures is laid down as a rule by western part of the world and the system is called International system of Unity (SI)

When science cannot accept anything which is not practically observed and experienced then I am most modern because I don't accept death as I haven't experienced it.

Later on when I read Bhagwad Gita my awareness have increased I was aware about my deep inner awareness of my body and I.

When I went deep inside my inner being I could experienced I am not the body

12 30 afternoon at docks, both of us are relaxing after lunch, naren and I, naren went to sleep, I am sitting in sukhasana, folded legs, naren sleeping on my lap, I am seeing the light behind my closed eyes and my body started growing, went on growing enveloping the vast universe, naren sleeping on my lap even in such a strange experience, I remember the talks done by naren the questions asked by him to me and our talks about family and environment and society.

This sleeping naren, now with his individual identity, will my consciousness affect his. Yes definitely it will...

Travelling in "vijaydoot" the name of a product carrier ship, when we entered at Bengal harbor at Netaji Subhash dock going through the river seeing the greenness and people over there, years had passed after British invasion and industrial revolution but the greenness still remains, children's playing, small huts people living in, still the beauty remains. The soul is there.

But there in France when we were heading to Rouen, going from the river the same conditions prevailing, lands water greenery but more sophisticated. At France I was thinking with all modern amenities there is something missing later on I realized it was soul light of soul

Due to the effects of wars and invasions resulting poverty the houses are small huts in Calcutta harbor, but advanced houses or huts with glassed doors the cars parked outside everything looking attractive at France.

As we enter inside the houses became more prominent, the technology was superb, but at what cost of humanity and colonial extraction

One side of the world had lost to poverty the other side is growing; it takes me back to the time when people from distant part used to come to India to learn at NALANDA and various books had been written on appreciation on the technology.

199

They were doing surgery also at Nalanda but only if there wasn't any other alternative, they believed in being with the nature, using natural herbs as medicine preparing it in natural way and being aware about all the herbs and plants and trees, talking with them, there were mantras being used at the time of breaking leaves or even wood from the trees, they used to request the plant and pray that we are taking it for the medicine or for prayer.

There were even times fixed for breaking leaves one has to go through the descriptive calculated calendars as per astrology and astronomy. I used to discuss these things with my friends unni, Mathews, thawal and my Chinese friend Yu Chu.

At times within conversation unit's eyes goes glassy but still he is listening to me. I know but some other thought is also interrupting his concentration and I know it, and know he also knows it, without my saying, we both smile at each other.

At times when the talk is going about star and delta I don't know why I want not to understand it. I want something else something which is the very basis of electricity or star delta, I was satisfied about that something I was in search of while meeting Salini Desai.

Today there is hardly any place that does not make use of the results of Electricity.

Every (atom) anu emits rays of varying wavelengths; the whole universe is full of this anus.

The Sanskrit word anu or atom as given by the rishi kanada, can be translated as the atom or an undivided part or indivisible.

Electricity is a mysterious force we cannot see it as we see sun moon and other stars. We are aware of it and its results. We have to understand it. To do so we have to experimentally learn for example by moving the copper coil between the magnets. When we see the copper coil moving, we know the force is there. It has caused coil to move. It doesn't matter what theory you have gone through. It is like Gandhi ji taking ideas from Russian philosopher Kropotkin rather than Charles Darwin for making his eco-friendly philosophy of sarvodaya.

From past studies by philosophers like Kanada we know that everything constitute of Anus or atoms our bodies, star, air, water everything is made of atoms. They are building blocks of the creation, very tiny particles. If you concentrate between your eyebrows even with open eyes you can see whole space is full of innumerable particles. Atom is the minutest particle by which matter is composed. Matter can be defined as anything that contains mass and occupies some space

An atom looks like the sun with the planets spinning around it. The center point is called as nucleus. It includes protons and neutrons. Electrons move around nucleus in spherical formed energy levels. In a balanced atom there are same no of proton and electrons. It can have different number of neutrons. Neutrons are electrically neutral.

Protons are positively charged particles and electrons negatively. Opposite charges attract each other. In some of the energy levels electrons are closer to nucleus and in some they are far. These far ones can be moved. When they move their motion is called electricity. Before the invention of radio a Yogi explained to his disciple about subtle laws by creating an incident the plot of cauliflower robbery. Space consists of particles bigger or smaller atom or star. They vibrate at different frequencies these frequencies can be caught as done by radio and television. When you switch on the television and tune to the frequency with certain wavelength you see the desired three dimensional realities. But if you change the channel or station your perception will change

The plot of cauliflower robbery, from AUTOCIOGRAPHY OF A YOGI in order to teach to Paramhansa Yoganandaji, Yukteshwar giri directed his thoughtrons or thought waves to one person, hankcring for a cauliflower, same way as we tune the radio to correct in hz or frequency and get a desired radio station, and thereby by his will he directed the person to the spot where the cauliflower had been kept unprotected, the person opened the door, he could only take cauliflower, no other valuable things.

Yukteswarji told Yoganandaji , you would understand these subtle laws unraveled by modern scientists Soon after that scientific discovery was done. Radio was devised in1939.

Man himself as well as all kinds of non living matters constantly emit rays which was again confirmed by India's great scientist Jagdish Chandra Bose, he broke the barrier between the living and non living entities, confirming by the practical demonstration that they all vibrate at the set pattern

Sir Jagdish Chandra basu--- proved all particles as living this was done before independence so Jagdish Chandra Bose didn't became famous around the globe but it hardly matters he enjoyed every moment of his living working for global love for every particle.

He even proved it by applying chloroform to various plants and even tin, with the help of the instrument he devised that is "kriskograph" that how the vibration changes takes place.

All particles in the cosmos are lively it has been understood years back that few materials attract each other and few repel say for instance a magnet north poles of two magnets repel each other north and south attract each other for example if you rub a piece of silk cloth and glass rod you can feel the attraction between them. If this force is controlled and stored it can be used for other purposes, that's how the understanding of static electricity comes unto being. . In Atharveveda the hymns for the cure of diseases and of possessions by demons of diseases are given elaborately. Study of age prolongation and preservation of youth is taken as a special subject under the term "ayusjyan" which gave birth to another term "rasayana" the Sanskrit equivalent of alchemy. "The gold which is born from fire, the immortal they bestowed upon the mortal. Our systems are not scientific they are beyond the realm of science as metaphysics is beyond physics. The very basis of science is shaky which was noticed at Vienna circle. The value of science is due to application and new gadgets invented in aid to application. We can take help from this keeping awareness for eco-friendliness.

The origin of alchemical notions gathered around gold, lead, soma juice, medicinal plants even mud from different terrains. White gold was regarded as the elixir of life. The remnants of making of gold can still be traced in the books. In the works of Shri Ramakrishna paramhansa such sadhus descriptions are there, they used to turn metals into gold for their basic needs so they don't have to depend on society. Experiments have revealed that all objects are composed of extremely small particles known as basic building blocks known as atoms. And these atoms are further composed of particles known as protons electrons and neutrons, all atoms don't have neutrons. There isn't a statement that this is it and there won't be any change after that. Everything is in flux or constant change the seemingly potential state is kinetic and vice versa there will always be enough space for change for all. For example element of carbon consists of six protons six neutrons and six electrons, protons and neutrons are tightly bound at the center of the atom known as nucleus, quantity of protons at nucleus determines the identity of the element. It is said if you remove three protons from the nucleus of an atom of lead you can produce an atom of gold. Neutrons are hard to remove from the nucleus and they don't affect much to the identity of the atom than proton

Air waves of energy to which the human ear is sensitive is sound, when variable energy is applied to a fluid, wave form is created on the fluid and it passes it on to the succeeding layers of the fluid.

In this way the varying energy can travel through a fluid from one point to another.

Sound waves are converted into electric waves by means of a microphone; electrical waves are connected into air or sound waves by applying varying current to the loud speaker.

The waves or cycles per seconds are called as frequency. Frequency between 20 and 20000 cycles per seconds can be heard so they are known as Audio frequency.

Frequency above 20000 cycles cannot be heard and is called Radio frequency; similarly frequency below 20 cycles also cannot be heard.

At lunchtime at naval docks we use to sit at the shore, under the trees after lunch, enjoying the breeze and waves rolling, the sound and the silences.

Naren use to ask questions starting from technical changing into philosophical Enquirer---

Other times at evenings, at navy nagar there at sea shore again we used to have long talks till the time it was dark---

The question is what's the best and what's new both are part of circle, today what is new may be after some time becomes old what is important and interesting is time.

He asks what time is. Now this is the profound question, if he asks what is the time I can say 5 30 or any other, but it is what is time? Profound question Neither viewpoint is unusual, naren and I talking about family how the family word came into existence, why brother and sister, I could recollect from my memory about Yoganandaji saying why we come in this human form in the same family as brothers and sisters, he cannot understand many things because neither he has experience nor read about, but I like him because he talks in different issues of life and he is very calm.

Nice breeze is moving and the sun is slowly setting, we come here regularly to watch the scenic beauty of sunset, when I was in school I use to come alone here and sit for a long time, even after sun set because every moment is different and enjoyable. When wind becomes chilly it's time to go back home. My father always says, you should watch rising sun, not the setting, but his ideas are based on talks of his elders, it doesn't come from his own inner being, I love sun set.

Conversation doesn't go all the time, at time it is silence and we watch the nature and at other times at naval dock I think about this silence also-what it is

When Gagangiri Maharaj went in one of the old caves with his close associates, he saw a yogi sitting in Samadhi, he waited for yogi to be aware, after some time yogi opened his eyes and asked him, did divine mother came back after killing mahisasura, because I went in Samadhi after she went, and now I am arising from that. This silence from which the Yogi aroused was beyond conventional space and time.

When we speak or utter we apply a physical energy which is varying, up and down in a form of a wave, as we see in ocean, Gagangiri Maharaj was using a similar kind of technique to remain inside water for several months he used to set himself with these waves in such a way that he could breath and his meditation goes on Dr Ram Bhosle was another great man who could do this. He used to go inside water at beach on full moon days and meditate whole night. He was a master of many arts. These waves can be called as air waves since they move in an air as a medium to which human ear has an awareness, these air waves of energies can be called as sound.

When we talk we apply a physical energy to the airwaves, which is in a waveform or in a varying state, which can travel from one space to another.

When we speak in front of a microphone, a device by means of which sound waves are converted into electrical waves and therefore by applying currents to the loud speaker electrical waves are converted into airwaves or sound waves.

A wave in motion has certain behavioral pattern---

The no of such waves or cycles are termed as frequency. Air waves or frequency between 20 and 20000 cycles per second are known as audio frequency because they can be heard, below 20 cycles and above 20000 cycles cannot be heard.

Frequencies above 20000 cycles are known as Radio frequency.

Electrical energy can also create magnetic energy-------refer

Electromagnetic waves are produced by means of varying electric current. The velocity of the waves of any frequency is same as of the light waves.

Sound waves can be converted into electromagnetic waves, these electromagnetic waves can travel through space without the means of wires and it can be transmitted, depending upon the high and low frequency, high frequency waves can reach to farther distances.

Audio frequency waves are mixed with radio frequency waves, the process is defined as embossing the impression of audio frequency waves on radio frequency waves and known as Modulation.

These modulated waves are transmitted from the radio station. The points at which the wave is maximally displaced are uniform in radio frequency waves, but this uniformity varies when they are modulated Amplitude modulation AM is where amplitude changes, frequency remains constant. Frequency modulation, is where frequency changes amplitude remains constant?

The dial of a radio, Frequency of the RF used for transmission ranges from 500 kHz to 25 MHz ranges are calibrated on as the dial is moved to the required frequency, the radio frequency signals of the frequency will be received and reproduced as sound.

As we see in tape recorders, when recording is done, the individual speaks in front of a microphone, sound waves are converted into electrical signals by a microphone, electrical signals are amplified and these amplified audio frequency signals are fed to a magnetic head it converts these signals into magnetic signals and records them on the tape.

In a blank tape all the particles are disarranged, it is a plastic ribbon with powdered iron oxide coating set in a hard emulsion and recording means these disarranged particles are arranged in a form or a pattern.

As the plastic ribbon passes through the sound channel and meets the erasing head, then the previous recording gets erased.

When it passes through the recording head the ribbon particles are affected by the varying electric current at the head and the particles are arranged in a pattern and retained.

At play mode the ribbon passes along the play head, affecting the electric current, passed through the amplifier unit of the recorder and sound waves are heard.

A vision from a distant place through transmission is television. The audio waves are transmitted by modulation with radio frequency carrier waves in radio transmission same way vision or picture waves are transmitted by modulation along with radio frequency carrier waves, these are vision signals or video signals. These signals can also be given by advanced yogis

Swami Rama from Himalayas made his ashram at Pennsylvania demonstrated regulation of his body mechanism by showing his palm and making it warm and cold by his will.

When we apply perfume to our skin it gets a cold sensation, as it evaporates rapidly, taking heat away, and the similar effect happens in a lesser way when perspiration evaporates. We feel these sensations because the regulation is done seemingly involuntarily by the system. Same is done by the yogi voluntarily. Yogis believe in living with the nature. When we use air conditioner we may affect the environment adversely. The regulation of the heat in modern machine is human brain invention.

This happens because the heat absorbed by the liquid is turned into vapor. When we boil water in a open pot, the liquid is given to the liquid or water is absorbed by the liquid, we have raised the temp of the liquid, evaporation starts happening, when more heat is applied, the rate of evaporation increases, the temp rises till a point of the liquid is reached from there no further effect on the temp will take place, all heat will turn the liquid into vapor.

This heat absorbed by the liquid wherein, it turns into vapor is a hidden heat. If these vapors are condensed back to liquid, the hidden heat is released.

In A/C warm air passes through the evaporators or cooling coils, where it gives up its heat to the refrigerant which is circulating through the system, after this refrigerant has done the cooling by evaporating, the gas is collected for re liquefaction by using a compressor to suck gas from evaporators at low pressure and deliver it to a condenser.

This liquid from the condenser passes through an expansion valve, which regulates the flow of liquid to the evaporators.

The circuit from expansion valve to the suction valve of the compressor is called the low pressure side and from the compressor delivery valve to expansion valve is called the high pressure side, this pressure on both the sides are maintained by expansion valve. This is the mechanical way or the way of removing the heat and making the system cool you can use the conventional language and name it thermo because its heat and the subject thermodynamics or part of thermodynamics. In electricity it is flow of electrons in the same way pressures are there in electrical circuits also which is known as potential difference

In electrical understanding the wirings done are made in the form of logical sequence which is known as circuit diagrams

The use of a circuit diagram is to enable the reader to understand the operation of the circuit, to follow each sequence in the operation from the moment of initiating the operation (e.g. by pressing a start button) to the final act (e.9.starting of the motor). If the equipment fails to operate correctly, the reader can follow the sequence of operations until he comes to the operation that has failed. The components involved in that faulty operation can then be examined to locate the suspect item. There is no need to examine other components that are known to function correctly and have no influence on the fault, and the work is simplified. A circuit diagram is an essential tool for fault finding.

A wiring diagram shows the detailed connections between components or items of equipment, and in some cases the routing of these connections.

An equipment wiring diagram shows the components in their approximate positions occupied within the actual enclosure. The component may be shown

A Simple starter

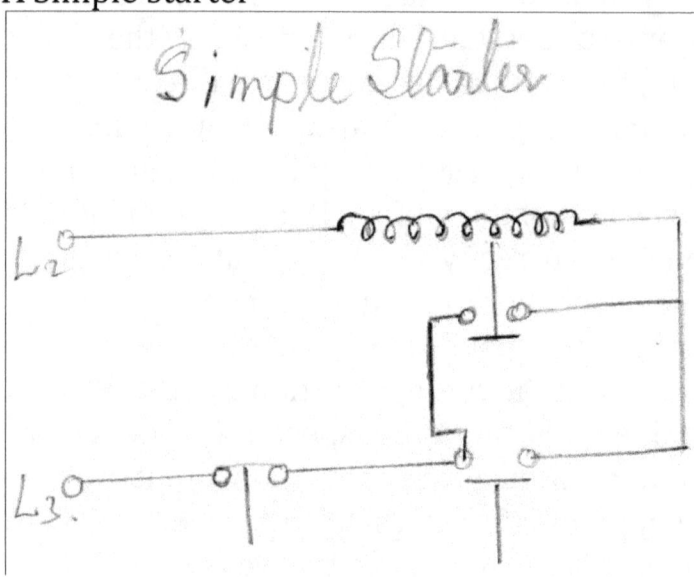

Start and stop push buttons on a starter are generally connected or wired in series, except start push button by construction is normally in break position.

When you depress it , it will make the contact whereas stop push button is normally in make position

When you depress it, it breaks the contact & the circuit is open ckt

Given above is a simple diagram after starting the push button which is generally in normally open state the coil is energized and auxiliary contact is made and now the circuit is completed through auxiliary contact and stop push button which is normally in a closed or make state. When stop push button is pressed it breaks the circuit coil is de energized and starter stops and hence the motor

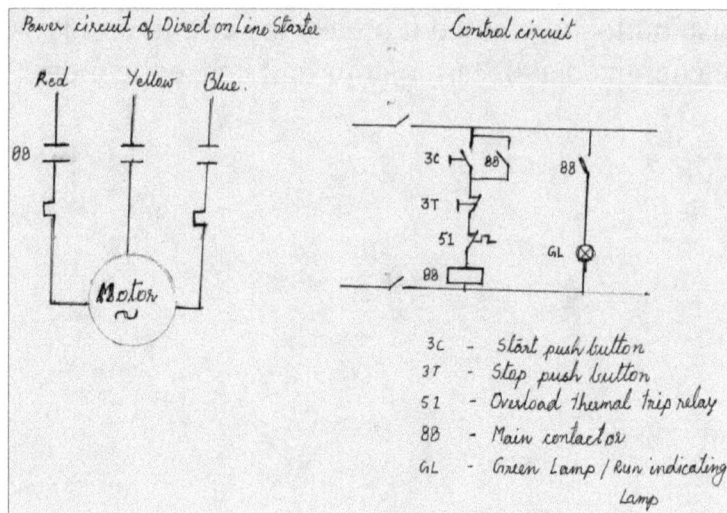

Power circuit of Direct on Line Starter   Control circuit

Red   Yellow   Blue.

88

Motor

3C - Start push button
3T - Stop push button
51 - Overload thermal trip relay
88 - Main contactor
GL - Green Lamp / Run indicating
                                    Lamp

diag

When start push button is depressed the coil gets the supply & hence it is energized, further making the auxiliary contact, & when the start push button is released the circuit is completed through auxiliary contact and stop push button. The motor starts, now when stop push button is pressed it breaks the circuit, coil is de energized and starter stops, further stopping the motor.

STAR DELTA STARTER

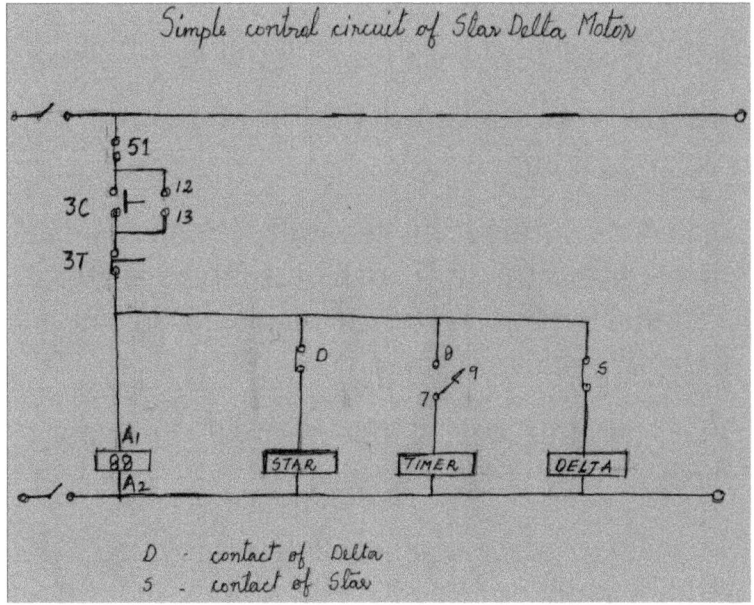

Simple control circuit of Star Delta Motor

diag

The right side is power circuit, wherein three phase supply is coming through No fuse breaker contacts and overload coils or over current coils to the motor. The left side is given with the control circuit. The supply for control circuit is taken from R & Y phases here, through the step down transformer from 440v to 220v it can also be further reduced to the required voltage.

Generally 3c term is given for start push button. At the left side corner

The first normally open contact made is 3c i.e. start push button in parallel

To 13 & 14 is provided, which is also a

Normally open contact, is a retaining contact.

In series with start push button, stop push button 3t is connected.

Again this 3t term will generally be found on all the diagrams for stop push button, thereafter it is connected to 95 & 96 normally closed contact of the over current relay.

Over current relay has one normally open and one normally closed contact.

Normally contact is generally given by 95, 96 and normally open as 97, 98.

Now it is connected to a1 of the coil and a 2 is connected to other side of the circuit .

Normally open i.e 97,98 of the over current relay is connected to the lamp for indication of overload trip

if it takes place, because if it happens the normally open contact will close and normally close will open .

The retaining contact 13, 14 or k is connected to another lamp for indicating motor is in running.

As we have seen how the control circuit is made by taking the supply from two phases through transformer.
These are practical tangible effects of scientific discoveries initiated by thinkers philosophers as you see great Einstein was influenced by Kant science will go on unless they come out with a material by which all are made. Please see my channel for further awareness on philosophy and practicality http://youtube.com/c/DrChandraSkeekhar

for videos on Tao of electricity i am giving here with the related diagrams

\

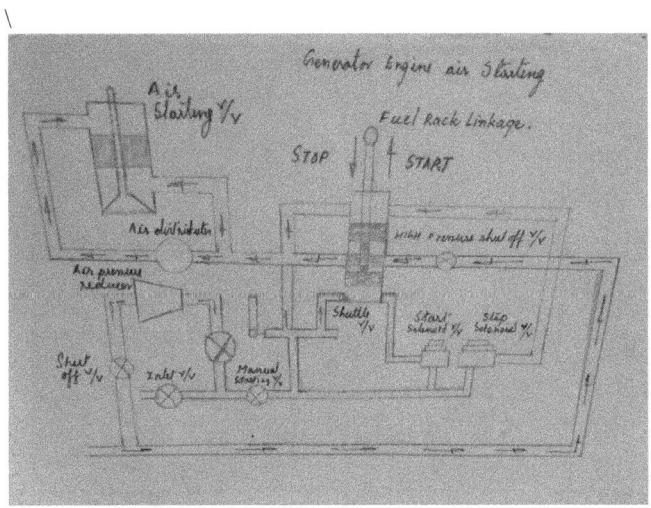

https://youtu.be/QK-_A4C622g

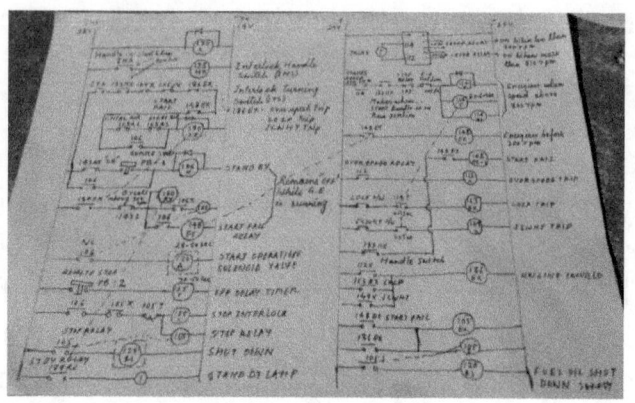

https://youtu.be/U-iSW5WWQ1s

Dr Chandra Shekhar names his art Vajramukti which means action to liberation- action in terms of using the techniques of Yoga and Martial arts that are means to an end...that of controlling mind and body, enhancing discipline and nonviolence.

Indian Express The prestigious Indian newspaper The modern Indian master Chandra Shekhar Bhatt is an exponent of a hybrid of martial arts and Yoga known as Vajramukti. He has had enough of a following to publish books, but you would have to go all the way to Bombay to train with him.

Shaolin society United Kingdom

Dr Chandra Shekhar has named this art Vajramukti which means action to liberation-action in terms of using the techniques of Yoga and martial art that are means to an end- that of controlling mind and body enhancing discipline and nonviolence This fusion has a common objective i.e having a higher level of awareness in life even while searching for absolute truth.

Chronicle Pharmabiz Mumbai

Dr Ram was very keen for the work on self enquiry and electricity to be published. He used to say Chandrasekhar actually first aero plane was tested at this land of Marathas on the basis of vimanshastra of Bharadvaja muni but the man who did had disappeared.

Dr Chandra Shekhar your quanta's of philosophical thoughts and in-depth knowledge of yoga & martial arts and electricity has to come in the form of philosophical enquiry who am I.

Dr RAM Bhosle student of Sir Herbert Barker United Kingdom

www.ingramcontent.com/pod-product-compliance
Lightning Source LLC
Chambersburg PA
CBHW060834170526
45158CB00001B/162